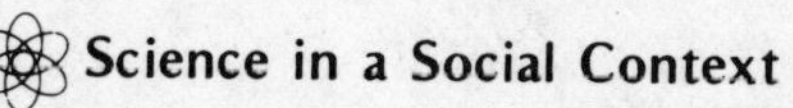

Assessment of Technical Decisions – Case Studies

Ernest Braun

and

David Collingridge
Technology Policy Unit,
University of Aston in Birmingham

Kate Hinton
Formerly Technology and Society,
Middlesex Polytechnic

Butterworths
LONDON - BOSTON
Sydney - Wellington - Durban - Toronto

The Butterworth Group

United Kingdom	Butterworth & Co (Publishers) Ltd
London	88 Kingsway, WC2B 6AB
Australia	Butterworths Pty Ltd
Sydney	586 Pacific Highway, Chatswood, NSW 2067 Also at Melbourne, Brisbane, Adelaide and Perth
Canada	Butterworth & Co (Canada) Ltd
Toronto	2265 Midland Avenue, Scarborough, Ontario, M1P 4S1
New Zealand	Butterworths of New Zealand Ltd
Wellington	T & W Young Building, 77–85 Customhouse Quay, 1, CPO Box 472
South Africa	Butterworth & Co (South Africa) (Pty) Ltd
Durban	152–154 Gale Street
USA	Butterworth (Publishers) Inc
Boston	19 Cummings Park, Woburn, Mass. 01801

First published 1979
ISBN 0 408 71313 5

British Library Cataloguing in Publication Data

Braun, Ernest, b.1925
Assessment of technological decisions.
1. Science – Social aspects – Case studies
2. Technology – Social aspects – Case studies
I. Title
301.24'3'0722 Q175.5 78-41268

ISBN 0-408-71313-5

Typeset by Butterworth Litho Preparation Department
Printed in England by Billing and Sons Ltd,
Guildford and London

Introduction

The aim of this collection of brief case studies, or case study notes, is to learn something about how technological decisions are made, and how such decisions might be improved. The introduction of new technologies has always involved some degree of risk, but today the scale and complexity of technology means that such risks must be very carefully considered. So great is the damage which can arise from an ill-judged decision to adopt a technological innovation that such decisions should be as carefully considered as is humanly possible. Fears have been expressed about the consequences of each of the technologies we have chosen to consider. These range from the unmitigated disaster of thalidomide to the controversial blessings of supersonic transport.

If we were to draw up a balance sheet of benefits and disbenefits for any technology we would generally expect to find a considerable positive balance. Because we generally expect and obtain such a positive balance, the present collection concentrates on cases where the balance has arguably tilted the wrong way: with negative effects dominating over the gains.

Negative consequences of technologies can arise in two main ways: *either* they are unforeseen, *or* they are regarded as a tolerable price to pay for the advantages obtained or expected. If the negative consequences of a technology were not foreseen at the decision-making stage, we may ask the question why. Were they unforeseeable, or merely overlooked? If the negative consequences were foreseen and regarded as acceptable, complex questions of value judgments and interest groups arise.

In the introduction of any new technology there are two main phases in which decisions are made. The first phase is concerned with the decision to attempt the innovation. In this phase the product-champion within an organization must convince the decision-makers that the new product or process is promising, and they make the decision to develop it and to attempt to market it. In this phase it is relatively easy to define the decision-makers, though even here a fairly large number of people may be involved. The next phase is the one at which an invention can turn into an innovation: the marketing stage. Here the decision to buy or not to buy can be highly centralized or completely diffuse. There is only one possible purchaser for the Aswan High Dam; there are millions of individual purchasers for D.D.T.

Problems of forecasting consequences of innovations are always formidable and differ according to whether the decisions are diffuse or centralized. Even in the diffuse case, however, some degree of government control, and therefore assessment, is usually exercised. In this series of case studies the reader should ask several questions about assessments. Were all the consequences foreseen and if not, why not? Were they unforeseeable and if so, why? What attempts were made to assess all the consequences of the technology? Who made the assessments and at what stage of the process of innovation? Were alternative projects considered?

Each case study consists of a brief account of the 'story', written to highlight the problems arising from the case. A bibliography is also provided to supplement the sparse text. For any serious study it is essential to read at least some of the books and articles listed.

Each study also contains a list of topics and issues which can be pursued in further reading and discussion. These issues are specific to the cases. In addition, there are several general issues, briefly hinted at in this introduction, which should be borne in mind. In a brief summary, they might read as follows.

The issues

1. Which of the undesirable consequences of the technology were foreseen, and what importance was attached to them by the decision-makers in the various stages of the innovation?
2. Why were there any unforeseen consequences?
3. Were some of the undesired consequences a result of upsetting the balance in some otherwise smoothly operating system?
4. What beliefs and value-judgments lay behind the decisions?
5. What were the stages in the decision-making process and who made the decision at each stage?
6. Was the decision to go ahead with the particular technology, on balance, 'good' or 'bad'?
7. How could the decision-making process have been improved so as to increase the chance of a 'good' decision emerging?
8. Were the decisions made in such a way that they could be reversed if unexpected, undesirable consequences were discovered?
9. Were alternative ways of achieving the desired goal considered?

Some of these questions, and the way in which they are linked, are shown in *Figure 1*.

Figure 1 Analysing why a technology with damaging consequences was adopted

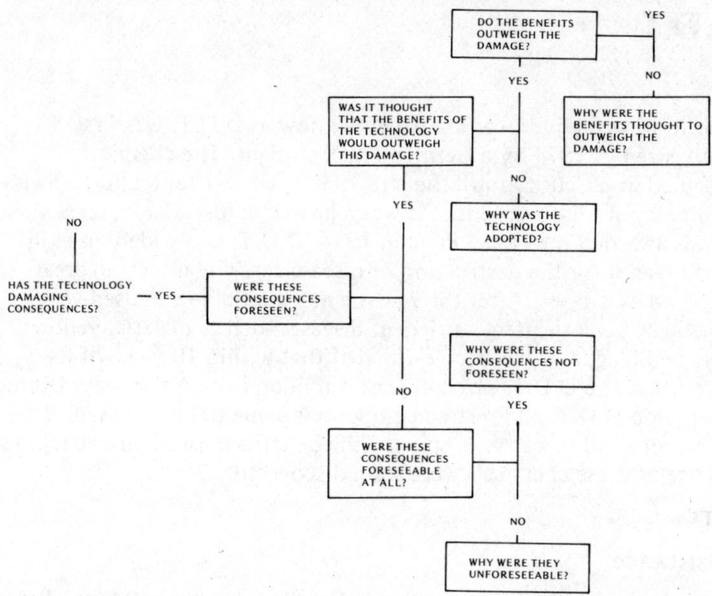

Concorde considers some of the problems inherent in a large project which represents a great technological advance, whilst *The Aswan High Dam* looks at the very difficult problems of balancing gains and losses for such a large scale interference with a natural system. *The Pollution of Water by Nutrients* is an example where centralized and diffuse decisions interact in very complex, and perhaps unpredictable, ways. *Thalidomide* is included as a classical example of a technological innovation with disastrous consequences, but it also raises serious questions about how such innovations should be tested for the possibility of damaging results. *D.D.T.* is also viewed by some as a classical case of harmful technology, although in reality its harmfulness is a matter of heated debate. It illustrates both the problem of predicting the consequences of a new technology and the problem of how to weigh benefit and loss. *Road and Rail Transport Policies* looks at, amongst other things, how accounting techniques can lead to a distorted view of the costs and benefits of a technological proposal, and hence bad decision-making. *Car Production Technology* deals with some of the forces which determine decisions about manufacturing processes, and the consequences which these decisions can have. *The Fast Breeder Reactor* is a technology which is seen by some experts as essential to the continued existence of the Western way of life, and by others, no less qualified, as an unnecessary and costly threat to human health and liberty.

Case Study One
D.D.T.

The chlorinated hydrocarbon we now know as D.D.T. was first synthesized in 1874 by a German Ph.D. student. The chemical remained in obscurity until the late 1930's, when Paul Muller, a Swiss entomologist, discovered that it was a powerful insecticide, for which he was awarded the Nobel Prize in 1944. D.D.T. was widely used in World War II for the destruction of disease-carrying insects in areas occupied by troops. After the war the new pesticide was used very effectively against insect carriers of diseases such as malaria, yellow fever and plague. It has been estimated that within 10 years of its widespread use D.D.T. saved at least 5 million lives in this way. During this period D.D.T. and related compounds came to be widely used in agriculture and forestry. In spite of the benefits derived, drawbacks to the use of these chemicals were soon discovered.

Resistance

Italian houseflies became resistant to D.D.T. as early as 1947, and soon resistant populations of lice, fleas and malarial mosquitoes were found. Twelve insect species were known to have populations resistant to D.D.T. in 1948, rising to 25 in 1954, 76 in 1957, 137 in 1960 and 165 in 1967.

Accumulation

D.D.T. becomes concentrated in the fat of organisms which come into contact with it. This enables the chemical to become concentrated in a food chain, a phenomenon known as 'biological magnification'. For instance, bottom mud in Green Bay, Wisconsin, was found to contain 0.014 parts/10^6 D.D.T., whilst the bay's crustaceans had 0.41 parts/10^6, fish 3–6 parts/10^6 and herring gulls, at the top of the food chain, 99 parts/10^6 of the chemical. This was enough to interfere with the gulls' breeding by causing weak, thin shelled eggs to be laid resulting in a low breeding success. Man has not escaped from this effect and people in the United States have averaged 8–10 parts/10^6 D.D.T. in their body fat, although there is no evidence that such concentrations are harmful.

Spread

Scientists were surprised to discover traces of D.D.T. in the penguins of Antarctica, thousands of miles from the nearest area of pesticide application. It appears that some D.D.T. finds its way into the sea where it is concentrated in food chains in just the same way as on land.

Persistence

The half-life of D.D.T. in the soil is around 12–15 years, and so concentrations can build up over the years in areas where it is used regularly. Some well sprayed orchards have accumulated as much as 113/lb of D.D.T. per acre after 6 years, retaining up to 40% of applied D.D.T.

Non-selectiveness

The red spider mite has been known on outdoor fruit trees for many years, but only achieved the status of a pest when orchards were sprayed with D.D.T. The mite is fairly resistant to D.D.T., unlike its predators which are quickly killed off. Such unexpected results of pesticide application are all too common because these chemicals are toxic to a very wide range of insects. Often many of the pest's predators are killed along with the pest, and, as the predators will generally be slower breeding than the pest, reinfestation quickly follows spraying.

These problems have led to the banning of D.D.T. and related compounds in many countries, and to a search for better methods of pest control, such as the use of less persistent more selective chemicals, the introduction of predators, the planting of sterilized pests, the development of resistant crops and improved cultural methods, and the use of insect hormones and attractants. The amount of research done and money spent on non-chemical control is, however, as yet nowhere near that involved in the search for new chemical pesticides.

The issues

The case study may be used to explore, amongst others, the following questions.

1. Should D.D.T. have been introduced more slowly so that its effects could have been better monitored?
2. Is enough spent on alternatives to chemical control of pests? If not, why not?
3. Why was the so-called 'mass control' approach to pests adopted in the United States, e.g. against the fire ant?

4. Was the banning of D.D.T. in many countries a reasonable step, or an hysterical reaction to public opinion?
5. Were the adverse effects of D.D.T. foreseen and, if not, why not?
6. Why was chemical control of pests developed before large scale control by non-chemical methods?
7. Is it possible to assess the cost of the problems associated with D.D.T.? Can this be compared with the benefits derived from D.D.T.?
8. D.D.T. has saved human lives, but at the cost of damaging other species. Is it possible to compare the value of a human life with the well-being of some animal species?
9. What were the advantages of D.D.T. over earlier insecticides?
10. Should a new chemical insecticide only be used when its safety has been proved, or should it be assumed safe until proven harmful?
11. What is the chain of decisions which runs through the D.D.T. story, from the decision of the chemical companies concerned to test the chemical to the decision, in many countries, to ban it?

Bibliography

1 *Approved Products and their Uses for Farmers and Growers,* published annually by the U.K. Ministry of Agriculture, Fisheries and Food.
Gives a list of approved products for pest control, advice on their handling and application, and legal restrictions on their use.

2 Berry, J. *et al.* (1974). *Chemical Villains,* pp 147–153. St. Louis; Mosby
A brief review of the usual problems laid at the door of D.D.T., including the possible production of cancers and genetic damage.

3 Carson, R. (1962). *Silent Spring.* London; Hamish Hamilton
By greatly exaggerating the potential hazards of persistent pesticides such as D.D.T., this book performed a useful service in focusing more attention on the problem and opening it up to public debate. Whilst many of its claims must be treated with great caution, this work has a central place in the pesticide controversy.

4 Challis, E. (1974). 'The Approach of Industry to the Assessment of Environmental Hazards.' *Proc. R. Soc.* Series B, **185,** No. 1079, 183–198
How industry investigates the likely environmental impact of some of its proposed chemical products.

[5] Commoner, B. (1972). *The Closing Circle*, Chapt. 8, pp 140–177. London; Cape
Commoner's thesis is that serious pollution is the result of a qualitative change in technology which began around 1945. Today the same basic needs are supplied as then, but by technologies which use artificial materials in place of naturally occurring ones. These new technologies have, therefore, a much greater impact on the environment than the old ones. D.D.T. falls into this pattern.

[6] Conway, G. (1971). 'Better Methods of Pest Control.' In *Environment: Resources, Pollution and Society*, pp 302–325. Ed. by Murdoch, W. Stamford, Conn; Sinauer
Chemical pesticides such as D.D.T. pose hazards to wildlife and man, and they often fail to control pests. This article explores the connection between these two failings and reviews alternative methods of pest control now being developed. A useful bibliography to more technical work is also given.

[7] Edwards, C. (1970). *Persistent Pesticides in the Environment.* London; Butterworths
A useful review of the problems associated with persistent pesticides, including their occurrence in the physical environment and in the biota, and the possibility of their control. A large number of tables summarising the results of many individual papers are included, together with an extensive bibliography, mostly to the technical literature.

[8] Farvar, M. and Milton, J. (eds.). (1973). *The Careless Technology.* Tom Stacey Papers 21–24 and following discussion, pp 373–466
A series of papers read at the Conference on the Ecological Aspects of International Development in 1968, dealing with the environmental impact of chemical pesticides and alternative techniques of pest control, with numerous case studies and many useful references to the more technical literature.

[9] Goodman, G. (1974). 'How Do Chemical Substances Affect the Environment?' *Proc. R. Soc.* Series B, **185,** No. 1079, 127–148
An excellent account of the difficulties involved in determining whether a new chemical product will damage the environment.

[10] Gunn, D. (1972). 'Dilemmas in Conservation for Applied Biologists.' *Ann. appl. Biol.* **72,** 105–127
Maintains that many arguments against the use of D.D.T. and its relatives are mistaken, unscientific and hysterical. Despite this, these arguments are repeated time and time again. In reality D.D.T.'s benefits far outweigh the occasional, short term damage caused by its misuse. A very stimulating paper with many useful references.

[11] McL. Philp, J. (1974). 'A Multi-national Company, The Public and the Environment.' *Proc. R. Soc.* Series B, **185,** No. 1079, 199–208
How Unilever ensures that the marketing of its products will not represent a hazard to the consumer or damage the environment.

12 Maddox, J. (1972). *The Doomsday Syndrome,* pp 101–116. London; McGraw-Hill

A defence of the use of D.D.T., pointing to its benefits in public health and agriculture, in reply to R. Carson's *Silent Spring* above.

13 Mellanby, K. (1967). *Pesticides and Pollution,* pp 120–132, 143–160 and 186–202. London; Collins

A balanced, non-technical look at D.D.T. and related pesticides, considering the benefits from their use, their effect on animal life, spread in the environment and alternatives to their use.

14 Moore, N. W. (1964). 'Pesticides and the Environment.' *J. appl. Ecol.* Supplement 3, Cape, Holden and Edwards

One of the early papers raising concern over the effects of pesticides on wildlife and the environment generally.

15 O.E.C.D. (1971). *The Problems of Persistent Chemicals,* pp 19–94. Paris

A survey of how the countries of the O.E.C.D. have acted to control the use of persistent pesticides such as D.D.T. Now a little out of date.

16 *Pollution: Nuisance or Nemesis,* pp 15–20. London; H.M. Stationery Office

A brief account of the history of D.D.T. and its impact on the environment. No depth of treatment is attempted and no figures are given, but it may be useful for introductory material.

17 Rothman, H. (1972). *Murderous Providence,* pp 98–112. London; Rupert Hart Davies

A brief discussion of the pesticide problem including public reaction, government action in various countries, economic factors in the banning of D.D.T. and the neglect of their social responsibility by involved scientists. Alternative methods of pest control are also mentioned, with a telling comparison of R & D expenditures for chemical and for biological control.

18 Rudd, R. (1964). *Pesticides and the Living Landscape,* Madison; Univ. Wisconsin Press

A much more balanced and researched view of the problem than that provided by R. Carson's *Silent Spring* above. It is the author's opinion that careful use of chemical pesticides is an essential stop-gap until more selective methods of pest control are available. Many chemical treatments are, however, careless, unbalanced and damaging, so that work on alternative techniques is urgent. The book covers all aspects of pesticide use and the impact of these chemicals on the environment and on living organisms. An extensive bibliography, mostly to the technical literature, is given. Most of the examples discussed are from the United States (*see also* Rudd, 1971 below).

19 Rudd, R. (1971). 'Pesticides.' In *Environment: Resources, Pollution and Society,* pp 279–301. Ed. by Murdoch, W. Stamford, Conn.; Sinauer

A very useful overview of the pesticide problem. The origins, misuse, toxicity and persistence of D.D.T. are discussed. The basic problem is that chemical pesticides are unselective and difficult to control. The author concludes that these chemicals should be banned. A bibliography is given (*see also* Rudd, 1964 above).

20 Ingersoll, B. 'D.D.T. on Trial in Wisconsin'; Wurster, C. 'D.D.T. Goes on Trial in Madison' and Laycock, G. 'The Beginning of the End for D.D.T.' (1971). In *Understanding Environmental Pollution,* pp 56–80. Ed. by Strobbe, M. St. Louis; Mosby

Three articles on the debates which eventually led to a nationwide ban on the use of D.D.T. in the United States. Fascinating reading.

21 Study of Critical Environmental Problems (SCEP). (1970). *Man's Impact on the Global Environment,* pp 126–136. Cambridge, Mass.; M.I.T. Press

A detailed account of D.D.T. in the marine environment, recommending the future banning of D.D.T. and its replacement by biological methods of pest control. This may involve the paying of subsidies to underdeveloped countries.

22 Van Buskirk, R. (1971). 'Requiem for an Old Friend.' In *Understanding Environmental Pollution,* pp 52–55. Ed. by Strobbe, M. St. Louis; Mosby

A humorous, but well argued, defence of D.D.T. and attack on those who wished to ban it.

23 Wagner, R. (1974). *Environment and Man,* 2nd ed., Chapter 13, pp 266–288. New York; Norton

A useful, well written, survey of the problem, including a brief account of 'integrated' pest control.

24 Walker, C. (1971). *Environmental Pollution by Chemicals,* Chapter 5, pp 55–76. London; Hutchinson

A fairly technical account of most of the problems resulting from the use of D.D.T. and related chemicals, which could prove very useful.

25 World Health Organization, (1972). *Health Hazards of the Human Environment,* pp 205–209. Geneva; WHO

A brief review of the hazards to human health from D.D.T. and its relations, with a useful bibliography. The safety record of D.D.T. is very impressive.

Case Study Two
The Fast Breeder Reactor

Most of the cases discussed in this Unit will concern decisions which have already been made. The present case study, however, is about a decision which, at the time of writing (March, 1977), lies in the near future. The decision is whether the U.K. should begin a programme of full-scale commercial fast breeder reactors. The Minister of Energy, now Mr. Wedgwood Benn, is expected to announce his decision whether to build the first reactor of such a programme within the year. Both proponents and opponents of the fast breeder reactor see this decision as perhaps the most important single decision about technology which has ever faced the country. If the supporters of the reactor are right, and the Minister decides against the programme, there will be a severe shortage of energy in the early years of the next century which will affect everyone in the country. If the reactor's critics are correct, and the Minister decides to begin a programme of breeder reactors, then we will be saddled with a very expensive, useless, but highly dangerous technology, the protection of which will infringe the freedom of every citizen.

Reactors

THERMAL REACTORS

The technology of the fast breeder reactor is best seen in contrast to the working of an ordinary, or thermal, reactor. A thermal reactor produces energy from the breaking apart, or fission, of atoms of uranium. Not all uranium atoms undergo fission, however, but only those of the isotope uranium 235, which makes up about 0.7 per cent of naturally occurring uranium. When an atom of uranium 235 breaks up it releases energy and produces atoms of lighter elements, called fission products, and several neutrons. The neutrons are emitted at high speed, but if they are slowed sufficiently by being made to pass through water or graphite, called moderators, the neutrons can cause further atoms of uranium 235 to break up. Neutrons slowed in this way are called thermal neutrons, hence the name 'thermal' reactors. Consider an atom of uranium 235 in a reactor which breaks up to emit three fast neutrons. Two of these neutrons may be lost, by escaping from the reactor or by being absorbed by various atoms, leaving one neutron which, when slowed by a moderator, can bring about the fission of another uranium 235 atom. If this atom breaks up to give just enough neutrons to cause one more uranium 235 atom to undergo

Figure 2 Magnox reactor

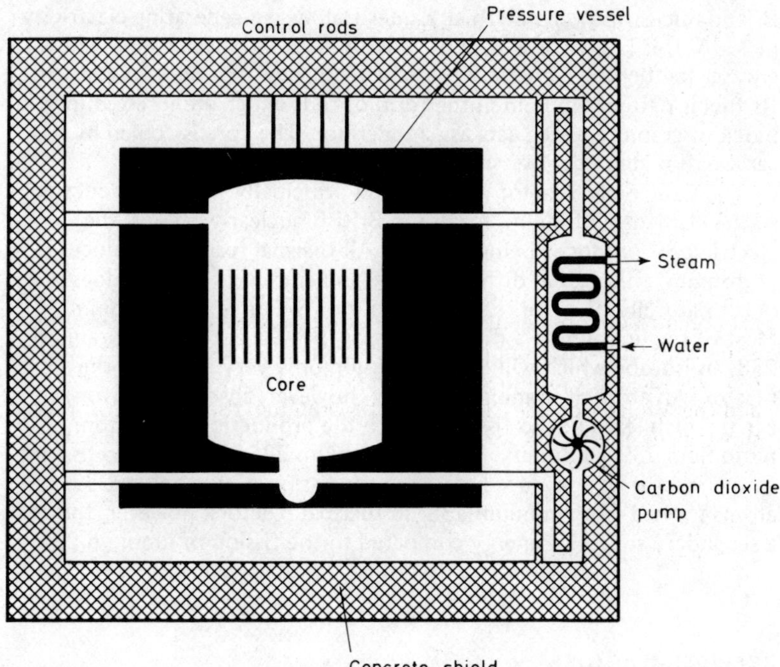

fission, then the fission reaction can continue indefinitely and the reactor is said to be 'critical'. If not enough neutrons are produced in the reactor to sustain the reaction, then it slows to a halt.

The number of neutrons in the reactor is adjusted by the movement of control rods which contain a powerful neutron absorber, such as boron. If the rods are lowered into the core of the reactor, where the fission reaction occurs, they absorb neutrons and slow the reaction. If they are raised, the number of neutrons in the core, and hence the rate of the fission reaction, increases.

Heat produced in the reactor core is generally removed by carbon dioxide or water, which is then passed through a heat exchanger to raise steam which is used to generate electricity in the conventional way. Problems of radioactivity leakage prevent the carbon dioxide or water heated in the core from being used directly for electricity generation in many designs of reactor.

A thermal reactor, therefore, has the following major components: uranium fuel, in the form of the metal or its oxide (which may be natural uranium or uranium enriched to 2–3 per cent uranium 235); a moderator to slow neutrons (ordinary water, heavy water or graphite); control rods which absorb excess neutrons (usually of boron steel); a coolant (carbon dioxide, ordinary water or heavy water) and a thick biological shield to prevent the escape of damaging radiation. The first British nuclear power station at Calder Hall began generating electricity in 1956. It is known as a Magnox reactor after the special alloy which encases its fuel rods. *Figure 2* is a schematic view of a Magnox reactor. Its fuel is natural uranium in the form of rods which are inserted into a block of graphite which acts as a moderator. The core is cooled by carbon dioxide under pressure.

The Calder Hall reactor was designed principally for the production of the element plutonium, for use in British nuclear weapons; the electricity it produced being a bonus. All thermal reactors produce plutonium, although at different rates, and since this element does not occur naturally, thermal reactors are our only source of plutonium. Most of the uranium in a thermal reactor (97–99 per cent) is uranium 238, an isotope which will undergo fission only very slowly in the reactor. An atom of uranium 238 may, however, absorb a neutron, an event which leads not to fission, but to the production of an atom of plutonium 239. Plutonium 239, like uranium 235, will undergo fission, so the reactor can produce heat from the fission of plutonium 239 atoms formed from uranium 238. In thermal reactors, however, this is a secondary source of energy compared to the fission of uranium 235.

BREEDER REACTORS

A breeder reactor is so called because as well as yielding energy it produces more fuel than it consumes. It is designed to produce new fuel from uranium 238 which absorbs neutrons in the reactor to become plutonium 239. Fast neutrons are best for this reaction and this leads to special problems in breeder design. To produce the many neutrons needed for the efficient conversion of uranium to plutonium a fission reaction is necessary. The conversion, however, requires fast neutrons while we have seen that to sustain the fission of uranium 235 atoms slow neutrons are needed. The fission of uranium 235 cannot, therefore, be used to produce fast neutrons for the conversion. The problem is solved by the use of plutonium 239, whose atoms will undergo fission when bombarded with fast neutrons. A breeder reactor is, therefore, ideally fuelled with plutonium 239. Fast neutrons from the fission of plutonium 239 atoms can produce fission in other plutonium 239 atoms, thus sustaining the fission reaction, and can also convert uranium 238 into plutonium 239. If the reactor is designed correctly, it will produce more plutonium 239 than is consumed in the fission reaction. When enough new plutonium 239 has been created, it can be used to fuel a new breeder reactor. Because they employ fast neutrons, breeder reactors are often referred to as fast breeders.

Figure 3 represents the British breeder reactor, which, for various technical reasons, is fuelled with a mixture of 80 per cent natural uranium oxide and 20 per cent plutonium 239 oxide. There is no need for a moderator, as this is only necessary to slow neutrons. In certain parts of the core is placed a uranium 238 'blanket' which is eventually

Figure 3 Breeder reactor

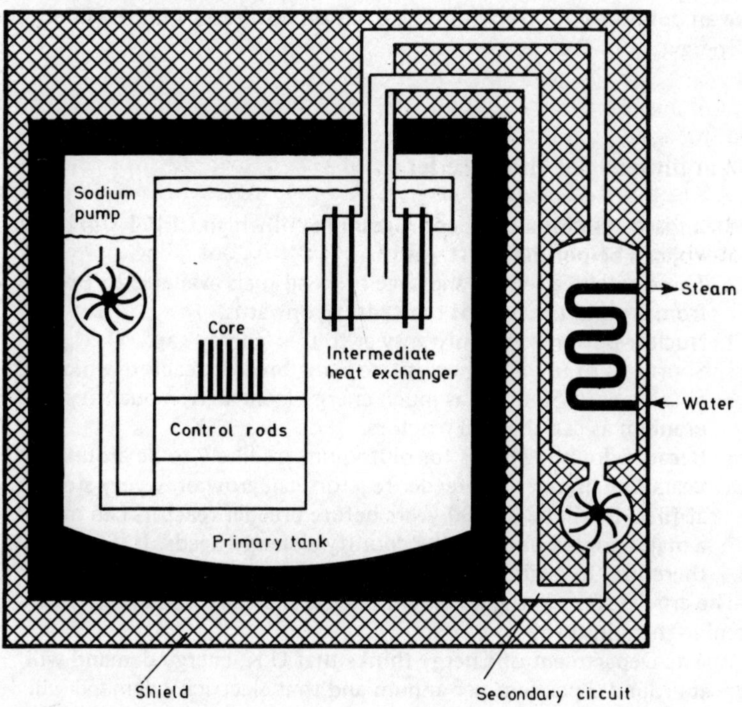

converted to plutonium. The core is very compact and must, therefore, be cooled very efficiently. This is done by means of liquid sodium. Because the sodium becomes radioactive, a primary and a secondary heat exchanger are needed. This design is sometimes called the liquid metal fast breeder reactor (LMFBR).

At first glance, the breeder reactor seems to break all the rules of thermodynamics, but all that happens is that neutrons from the fission reaction are used to release the energy locked up in uranium 238 atoms, by converting them to plutonium 239 atoms. The production of new fuel by present breeders is, however, not spectacularly fast. For every ton of plutonium in such a reactor there will be about one ton of plutonium in various stages of reprocessing and fabrication and in transit. This so-called 'pipeline inventory' is the total plutonium associated with one reactor. Breeders of present design are likely to double this quantity of plutonium, and so produce enough of the metal for an additional reactor, in about 20 years.

The first true power reactor to employ the breeding principle was the British Dounreay Fast Reactor which first generated electricity in 1959. The Prototype Fast Reactor also at Dounreay, producing 250MW of electricity, followed in 1975. It is now proposed to build the world's first fully commercial breeder reactor (CFR 1) which is expected to have an output of 1300MW(e). The decision now rests with the Minister of Energy.

The argument for the breeder

The standard case for a breeder programme, of which CFR 1 will be the first step, can be put in four stages:

1. There will be a serious shortage of fossil fuels available to Britain from around the turn of the century onwards.
2. Nuclear power is the only answer to this 'energy gap'.
3. Shortages of uranium necessitate using breeder reactors which can obtain about 50 times as much energy from a given quantity of uranium as can thermal reactors.
4. Because doubling times for plutonium are likely to be around 20 years, the number of breeder reactors can grow only very slowly at first. It will be 20–30 years before breeder reactors can make a major contribution to the country's energy needs. It is essential, therefore, to begin a breeder programme as soon as possible.

The crucial steps in the argument are the first three, so let us examine them in turn.

1. The Department of Energy thinks that U.K. energy demand will grow at around 2 per cent per annum and that electricity demand will grow at around 4 per cent per annum over the next 20 years. On this projection energy demand will double by the year 2010, and electricity demand could well double before the end of the century.

2. By this time North Sea oil will be nearing exhaustion, and the exhaustion of major foreign oil fields will prevent the import of oil. Coal is not expected to be available in anything like the quantity needed to replace oil.

> '. . . the trend of demand for electricity and doubts about the availability of fossil fuels (and other alternatives) on a sufficient scale point to the need for a large and increasing nuclear component in our energy supplies by the turn of the century'. (Department of Energy, *Submission to the Royal Commission on Environmental Pollution, Study of Radiological Safety*).

3. Known reserves of uranium will be exhausted by the nuclear reactors now in operation and those already planned. Although large areas of the world have not been prospected for uranium, and although thorium may represent a partial substitute for uranium, there is serious concern about the element's future availability.

> '. . . the United Kingdom has no important uranium resources of her own, and if it remained totally dependent on thermal reactors it would become increasingly vulnerable to the world price and availability of uranium. Fast reactors would reduce the impact of increases in uranium prices and reduce the possibility of our not being able to supply our demand for uranium from the world market. It seems essential, therefore, to keep open the option of using them'.
> (Department of Energy, *Submission to the Royal Commission on Environmental Pollution, Study of Radiological Safety*).

Objections to the breeder

DOUBTS ABOUT THE ENERGY GAP

Many have doubted the Department of Energy's projections of energy supply and demand. There are two basic objections: that energy demand will not grow as rapidly as predicted, and that non-nuclear energy sources will be able to provide a greater proportion of energy demand than predicted by the Department. Some argue that Government measures to conserve energy and to ensure its efficient use, for example, by discouraging the use of electricity for space heating and by using waste heat from power stations, can keep demand low enough to be satisfied by fossil fuels alone, or by fossil fuels supplemented by thermal reactors[5,7,8]. Others point to the possibility of exploiting novel energy sources such as waves, winds, tides, the sun and oil shale and tar sands[13,19,22].

URANIUM SUPPLIES

The issue of future uranium supplies is very tangled. Two quite contradictory arguments about these supplies and breeder reactors can be found. On the one hand, it is argued that there will be sufficient uranium to enable thermal reactors to meet British energy demands. The World Energy Conference's report, *Survey of Energy Resources*, argues that 'at costs up to $200 per kilogram the amounts of uranium available are in tens to hundreds of megatonnes and at costs up to $500 per kilogram in thousands of megatonnes'.

The mining of very low-grade ores is sure to pose considerable environmental problems and will yield expensive uranium. The cost of generating electricity is, however, not particularly sensitive to uranium costs. It has been estimated, for example, that a 10-fold increase in uranium prices will only double the cost of electricity generated from a steam generating heavy water reactor.

The outlook for uranium supplies looks even rosier when the work of Chapman[7] is considered. Doubts had been raised that producing uranium from low-grade ores would consume more energy than that obtainable from uranium. Chapman argues that even if the uranium is burned in thermal reactors this only happens for ores below 20 parts of uranium per million, a very low figure. It is not known how much ore exists down to such a low quality, but it is likely to be very large. It may be, therefore, that there is sufficient uranium to fuel thermal reactors without the need for breeders.

The contrary opinion has been expressed in the *6th Report of the Royal Commission on Environmental Pollution*[5], which suggests that uranium shortages may limit plutonium production by thermal reactors to such an extent that no large breeder programme will be possible.

REACTOR SAFETY

The most serious kind of reactor accident is the melting of fuel elements following overheating of the core. In a thermal reactor this is serious enough, but because the fuel is no longer in contact with a moderator the fission reaction will slow to a halt. In a breeder reactor, however, there is no need for a moderator to slow the neutrons and the molten fuel will become more reactive instead of less. If the reaction is severe enough to breach the reactor's containment, a very serious accident will result. There have already been two 'meltdown' accidents in the United States, but in neither was the containment broken. Although proponents of the breeder think that such accidents can be rendered virtually impossible by careful design, critics remain unconvinced.

REACTOR SITING

Doubts have been expressed about whether the U.K. contains enough sites suitable for the many breeder reactors needed to fill the Department of Energy's projected gap. One of the problems here is that safety considerations would seem to preclude the building of breeder reactors near urban areas. This also means that the vast quantities of waste heat which will be produced by so many reactors will not be available for space heating. On the Department of Energy's strategy more than half the country's total primary energy would be lost as waste heat from power stations in the year 2020 (*see 6th Report of the Royal Commission on Environmental Pollution*, paras 293–296).

PROTECTION OF RADIOACTIVE SUBSTANCES

Plutonium is one of the deadliest substances known. Nevertheless, when 40–50 per cent of the country's energy is from breeder reactors, transport of large quantities of plutonium will be an everyday event. Doubts have been expressed whether this could be done safely and whether the plutonium could be protected from terrorist groups without undue restrictions on civil liberties[29].

WASTE DISPOSAL

The waste products produced by all reactors, both thermal and fast, are of two kinds. Fission products are produced by the splitting apart of uranium 235 or plutonium 239 atoms, and activation products by the capture of neutrons by these atoms to form so-called higher actinides such as americium and technetium. Some fission products are very radioactive and require storage for a few hundred years before their radioactivity reaches safe levels. Activation products, however, require far longer for their activity to decline, some hundreds of thousands of years. A large nuclear power programme, using either thermal or fast reactors, will produce considerable quantities of waste and, as yet, no way of disposal is known to be safe for the very long time periods involved. The idea most favoured in Britain is for the wastes to be fused with a special glass, which will eventually be deposited in caverns dug in some extremely stable geological formation. British Nuclear Fuels is proposing to build a pilot plant for this glassification and is soon to begin prospecting for suitable geological sites in the U.K. Opponents of nuclear power, however, argue that these wastes are so long lived that their safe disposal cannot be guaranteed, and that the consequences of leakage could be very serious, both for the environment and for human health.

PROLIFERATION OF NUCLEAR WEAPONS

If technologically advanced countries like Britain produce breeder reactors, they will inevitably be exported to developing countries, just as thermal reactors are exported now. Breeder reactors, however, pose special problems because of the ease with which they can be used to manufacture nuclear weapons. Thermal reactors produce plutonium, but its recovery from spent fuel elements is a difficult task and much of the element is produced as the isotope 240. Plutonium containing a large proportion of plutonium 240 cannot be used to produce a powerful weapon of predictable yield. A breeder reactor's blanket, however, will contain plutonium 239, mixed with untransformed uranium 238 and fission and activation products. Although the ease with which the plutonium 239, a few kilogrammes of which are ideal for a nuclear weapon, can be separated should not be underestimated, it should not prove impossible for a country losing a war or desperate to impress its neighbours. Once the plutonium 239 has been separated, the construction of a weapon is relatively easy.

RESEARCH AND DEVELOPMENT EXPENDITURE

Many critics of the breeder reactor fear that its development will be so expensive that little will be left for research and development on other energy sources, so ensuring that the breeder is the only answer to future energy needs. Present expenditure would seem to support such fears. In 1973–74, for example, the U.K. spent a total of £94.4 million on all energy research and development of which £67.7 million was on nuclear research, including £32.9 million on the liquid metal fast breeder reactor.

The issues

The case study may be used to investigate, amongst others, the following issues.

1. How serious is the threat of an 'energy gap' in the U.K. in the early years of the next century?
2. What policies about conservation, fuel prices, research and development expenditure and education might reduce the threatened 'energy gap'?
3. What problems are likely to face British supplies of oil and coal in the early part of the next century?
4. Is there likely to be: (a) sufficient uranium to produce the plutonium needed in an expanding breeder programme; (b) sufficient uranium to enable all nuclear energy to be produced by thermal reactors without the need for breeders?
5. What criticisms have been made of the use of 'event trees' in assessing the safety of reactors? Are any of these criticisms valid?

6. What problems will the coming of the so-called plutonium economy pose?
7. Is enough research being done on alternatives to nuclear energy?
8. Can the wastes from nuclear reactors be disposed of safely, or will caring for them be a burden we place on future generations?
9. Is the spread of nuclear technology likely to lead to the proliferation of nuclear weapons? If it is, what can be done about it?
10. Will the use and transportation of nuclear material lead to intolerable restrictions on the citizen's civil rights?

Bibliography and references

BACKGROUND

[1] Department of Energy (1976). *Energy: The Key Resource,* Energy Paper No. 4. London; H.M. Stationery Office

[2] Department of Energy (1976). *Report of the National Energy Conference,* Energy Paper No. 13 (2 vols.). London; H.M. Stationery Office

[3] Hunt, S. (1974). *Fission, Fusion and the Energy Crisis.* Oxford; Pergamon

[4] Patterson, W. (1976). *Nuclear Power,* Harmondsworth; Penguin.

[5] The Royal Commission on Environmental Pollution, *6th Report: Nuclear Power and the Environment* (1976). Cmnd. 6618, Chapters 2, 3 and 4. London; H.M. Stationery Office

[6] Marion, J. (1974). *Energy in Perspective,* Chapter 5. New York and London; Academic Press

ENERGY FORECASTING

[7] Chapman, P. (1975). *Fuel's Paradise,* Chapters 9–12. Harmondsworth; Penguin

[8] Chapman, P. (1976). 'The all-electric dream.' *J. Br. Nuclear Energy Soc.,* **15**, 285–295

[9] Department of Energy (1975). *Evidence to the Royal Commission on Environmental Pollution – Study of Radiological Safety.* London; Department of Energy

[10] O.E.C.D. (1975). *Energy Prospects to 1985.* O.E.C.D.

[11] Roberts, P. and Outram, V. (1974). *A Method of Projecting Energy Demand in the U.K.* London; Department of the Environment
[5] Chapter 9

ALTERNATIVES TO NUCLEAR ENERGY

[12] Bromley, A. and Surrey, A. (1973). 'Energy resources.' In *Thinking About the Future.* Ed. by Cole, H. *et al.* London; Chatto and Windus
[13] Foley, G. (1976). *The Energy Question,* Pt. II. Harmondsworth; Penguin
[14] Department of Energy (1975). *UK Oil Shales: Past and Possible Future Exploitation.* Energy Paper No. 1. London; H.M. Stationery Office
[15] National Economic Development Office (1974). *Energy Conservation in the U.K.* London; H.M. Stationery Office
[16] Central Policy Review Staff (1974). *Energy Conservation.* London; H.M. Stationery Office
[17] Department of Energy, Advisory Council on Energy Conservation (1975). *Report to the Secretary of State for Energy.* Energy Paper 3. London; H.M. Stationery Office
[18] Parliamentary Select Committee on Science and Technology (1975). *1st Report, Energy Conservation.* London; H.M. Stationery Office
[19] Hill, P. and Vievoye, R. (1974). *Energy in Crisis,* Chapters 2–4, 8 and 9. London; Yeatmen
[20] King Hubbert, M. (1969). 'Energy Resources.' In *National Academy of Science, Resources and Man.* San Francisco; Freeman
[21] National Coal Board, *Annual Report and Accounts.*
[22] National Engineering Laboratory (1974). *Symposium on Re-assessing Energy Resources,* N.E.L.
[23] National Engineering Laboratory (1974). *R & D Routes to More Effective Utilization of Energy.* N.E.L.
[24] The Royal Society (1974). *Energy in the 1980's* (originally in *Philosophical Transactions of the Royal Society,* A276)
[5] Chapter 9
[9] Above
[10] Above

THE ARGUMENT FOR THE BREEDER

[25] Brookes, L. (1976). 'The plain man's case for nuclear energy.' *Atom,* April, 1976, 95–105

[26] Hill, J. (1975). 'The energy situation and the role of nuclear power.' *Atom,* January, 1975, 2–7
[27] Rotblat, J. (1977). *Nuclear Reactors – To Breed or Not to Breed.* London; Taylor and Francis
[9] Above

URANIUM SUPPLIES

[28] O.E.C.D., Nuclear Energy Agency and IAEA (1976). *Uranium Resources, Production and Demand.* (Reviewed in *Journal of the British Nuclear Energy Society,* July, 1976, 195–197 and *Atom,* May, 1976, 151–159)
[5] Paras 118–121, 455–457
[7] pp 102–109, 141–145

PROTECTION OF NUCLEAR MATERIAL

[29] *Nuclear Prospects* (1976). National Council for Civil Liberties
[30] Swindell, G. (1975). 'The safe transport of radioactive materials.' *International Atomic Energy Agency Bulletin* **17**, 1, 5–11
[5] Paras 182–185, 316–325

PROLIFERATION OF NUCLEAR WEAPONS

[31] Barnaby, F. (1974). *The Nuclear Age.* Stockholm International Peace Research Institute
[32] Barnaby, F. (1974). *Preventing Nuclear Weapon Proliferation.* Stockholm International Peace Research Institute
[33] *World Armaments and Disarmament, SIPRI Yearbook 1972,* Chapters 9 and 10. Stockholm International Peace Research Institute
[34] Collingridge, D. and Gutteridge, W. (1975). *The Arms Race and Problems of Disarmament,* SISCON unit
[5] Paras 182–186, 316–325, 330–333

RESEARCH AND DEVELOPMENT

[35] O.E.C.D. (1975). *Energy R & D.* O.E.C.D.
[36] Department of Energy (1976). *Energy R & D in the UK,* Energy Paper 11. London; H.M. Stationery Office. (Reviewed in *Atom,* July, 1976, 190–193.)

37 Price, G. and Eggington, T. (1976). Comments on[35]. *New Scient.*, 7th October, 1976, 32–35
38 Fells, J. and Hall, D. (1976). Comments on[35]. *New Scient.*, 29th July, 1976, 219–221
39 Chapman, P. and Conroy, C. (1976). Comments on[35]. *New Scient.*, 15th July, 1976, 121–124
5 Paras 509–515

SAFETY

40 Lister, B. (1975). 'Nuclear power – the perspective of risk.' *Atom*, May, 1975, 68–75
41 Farmer, F. (1975). 'Advances in the reliability assessment of reactor systems.' *Atom*, December, 1975, 218–226
42 Rasmussen, N. (1974). 'Nuclear power reactor risks in the US.' *Energy Policy*, December, 1974, 339–341
43 'Nuclear reactor safety – the issues' (1975). *Bull. atom. Scient.*, September, 1975
44 Lovins, A. *Nuclear Power*, pp 6–9 and Appendix 1. London; Friends of the Earth
4 Chapter 6
5 Chapter 6
13 pp 175–179

WASTE DISPOSAL

45 Dreschoff, G. (1974). 'International high level nuclear waste management.' *Bull. atom. Scient.*, January, 1974
46 Micklin, P. (1974). 'Environmental hazards of nuclear wastes.' *Bull. atom. Scient.*, April, 1974
47 Zeller, E. *et al.* (1974). 'Putting radioactive waste on ice.' *Bull. atom. Scient.*, January, 1974
4 pp 103–114
44 Appendix 2

Case Study Three
Concorde

The story starts in 1955 at the Royal Aircraft Establishment (RAE) with a working party to study the feasibility of civil supersonic transport (SST). Military supersonic aircraft were already flying, but the working party reached the conclusion that a civil supersonic airliner was not feasible as an economic proposition.

Within a few months the climate at RAE changed because of the emergence of a new aerodynamic idea. One or two scientists at RAE came up with a possible solution to the problem of the SST – a thin steeply swept wing, rather like a delta. Once the idea was there, the pressure began to build up. The scientists were raring to go, the manufacturers wanted the challenging work, and everybody wanted to get even with the Americans. A new high-powered working party was set up. This time it was representative of all aircraft interests in the country: Ministries, manufacturers, airlines, research establishments. The working party was Government financed and became known as the Supersonic Transport Aircraft Committee (STAC). It began work late in 1956 and reported in March, 1959. The report recommended the development of two supersonic aircraft, a long-range and a medium-range version. The estimated development costs were £50–70 million for the medium-range plane and £90 million for the long-range aircraft. The report was accepted by the Government and design studies were commissioned from aircraft manufacturers.

The medium-range version was soon dropped and the first design study for the larger aircraft was completed in the summer of 1961. In the meantime, however, the French aircraft manufacturer Sud Aviation announced their intention to build an SST and exhibited a model of the plane in June, 1961. At roughly the same time, the British Prime Minister, Harold Macmillan, told Parliament that the Government intended to apply for membership of the European Economic Community. The British Minister of Aviation and the French Minister of Transport soon arranged talks between Sud Aviation and the leading British contender in the SST field, the British Aircraft Corporation (BAC). More than a year of talks passed before a joint outline design was agreed in September, 1962. In fact at this stage there were again two versions of the design, a medium-range 220,000 lb 100 seat version and the long-range 262,000 lb 90 seat plane. The companies signed an agreement for the joint development of these two aeroplanes and shortly afterwards in November, 1962, the two Governments signed an agreement on sharing the cost of development. Britain's share of the total cost was then estimated at £75–80 million. The agreement contained no escape clause, that is, neither government could unilaterally withdraw from the project.

Four points need to be made at this stage of the story. (a) The development of an SST was a clear case of 'invention push', there certainly was no clamouring by airlines for such an aircraft. (b) The Anglo-French collaboration was politically inspired and Government sponsored, rather than entered into for commercial reasons by the plane-makers. The lack of an escape clause also appears to have had political reasons, possibly to increase the credibility of the will to co-operate. (c) The cost estimates were far too low. (d) Without Government finance the project would never even have got as far as the drawing board, let alone off it or into the air.

As soon as the agreement was signed, complex discussions on the detailed design work started. Gradually the plans for a medium-range version were dropped and the long-range version, Concorde, grew in capacity and weight. In March, 1964, the agreed weight was 326,000 lb and the aircraft could carry 118 passengers across the Atlantic. The Bristol Siddeley Olympus turbojet engine was to be re-designed to produce a thrust of 32,825 lb.

By this time the Americans had started their design studies for an SST and were planning to build a bigger and faster aeroplane than Concorde. By this time also several critical voices were beginning to make themselves heard. Was the total cost going to be ruinous? What about the sonic boom? What about hazards to health? And the noise? And was the whole thing worth while anyway? Shall we be beaten by the Americans yet again? The defeat of the Comet and even the VC10 by the first generation of American big jets still rankled.

In October, 1964, a Labour Government came to power. The economic position was precarious. The new Government had to find means of economizing on public expenditure. Amongst other things, all the major aircraft projects were to be reviewed. The review of Concorde was short-lived. It soon became clear that the French Government would not release the British Government from its obligations. The British options were therefore either to carry on or to be hauled before the International Court of Justice and to face a large bill for damages. The Government decided to carry on. Two other aeroplanes fell victim to the economic axe: the P-1154 and the HS 681. The TSR-2 was soon to follow them.

Despite the sales forecast of between 300 and 400 machines, delivery options were slow to come in. One of the reasons for this was the small capacity of the aircraft and in May, 1965, a further increase in size was decided. The aircraft was now to carry up to 136 passengers and its weight went up to about 350,000 lb. Delivery options slowly built up, to reach a peak of 74 in March, 1967.

By this time an anti-Concorde pressure group had been organized in Britain and the anti-SST lobby in the United States was gaining strength. The efforts of the latter were crowned by success in 1970, when the proposal to build an American SST was narrowly defeated in

Congress. Discounting the Soviet competition, this left Concorde as the sole SST in the Western world. Despite vociferous opposition and escalating cost the project went on. Estimates of growth in air traffic were optimistic, fuel costs worried nobody, and there was talk of a total sale of 500 Concordes in the early 1980s. When it became obvious that supersonic flying over land would not be tolerated because of the sonic boom, these estimates were halved, but remained at a respectable figure.

Several events proved decisive in making these estimates appear highly optimistic today, with only a handful of Concordes sold. In our view the most important of these events were as follows.
(a) The success of the jumbo-jets, whose operating costs are substantially lower than those for Concorde. (b) The reduced growth in air traffic, particularly in first-class business traffic. (c) The oil crisis, which makes the large fuel consumption of Concorde particularly unattractive. (Concorde uses roughly three times as much fuel per passenger-mile as a jumbo-jet.) (d) The general economic recession and the near-bankruptcy of almost all airlines. (e) The strength of opposition from environmentalists on many grounds: sonic boom, noise near airports, fuel consumption, cost, possible hazards to the ozone layer, low priority because the usefulness of an SST is limited to a small section of the community only, etc.

The present situation may be summed up in very few words: only nine Concordes have been sold, and even these only to the National British and French carriers with the promise of Government subsidies. The total development costs are now thought to be well over 1,000 million pounds and there is no hope of recovering these from sales. Indeed, it seems that each aircraft will have to be sold below cost price. The battle over landing rights in the United States is still on. Flights over land have been banned by most countries. There are few, if any, prospects of further sales. The manufacturers hope that the technical excellence of the aircraft, which few doubt, will eventually cause it to break through to success. The fate of Concorde is nearly, but not quite, sealed and opinion about it is as sharply divided as ever.

The issues

Amongst the many issues which arise out of this case, the following topics are suggested for discussion:

1. How serious could the impact on world oil reserves be if SSTs became universally accepted ([3], pages 106, 107; [4], page 233)? How do you reconcile these views?

2. Robin Maxwell-Hyslop[6] makes the following two statements.

> 'To many airlines, Concorde represents an opportunity to buy an aircraft they do not need, with money they have not got, to fly routes which may be forbidden to them. Such a temptation falls short of the irresistible.'
> 'For a country with few natural resources, it is particularly important to develop and maintain industries which achieve the highest possible ratio of export sales price to the cost of the imported raw materials. . . The aero-space industry exemplifies this, and Concorde exemplifies, however exotically, the highest form of that art.'

Can both arguments be simultaneously correct?

3. John Davis [3, Ch. 15] makes a strong economic case for Concorde. Using these arguments and those presented for example in [5], argue the case for and against Concorde in the years 1969 to 1971.
4. Argue the case for and against unilateral British cancellation of the project in 1964 and 1974. Consider also the case for Anglo-French agreement on cancellation.
5. Discuss the role of pressure groups throughout the history of Concorde.
6. Discuss the manner in which public control of expenditure was exercised.
7. Had the concept of Technology Assessment in its present form existed in 1958, would the Concorde project have been started? See particularly [1] page 23, describing the terms of reference of STAC. Do these seem adequate now?
8. What is the future for Concorde?
9. Is it true to say that Concorde is the result of pure invention-push? Are there no 'market' needs which it fulfills?
10. Discuss the problems of forecasting in relation to the Concorde project.

Bibliography and references

1 Davis, J. (1969). *The Concorde Affair.* Leslie Frewin
A book by an 'insider' of the aerospace establishment. Very

informative on technical aspects of design, performance, history and economics. Written with scrupulous objectivity by a committed person in the days when hopes for Concorde were high.

2 Blackall, T. E. (1969). *Concorde.* London; Foulis

A well produced book of detailed but intelligible description of the aeroplane. The most technical and least political/economic of the books.

3 Wilson, A. (1973). *The Concorde Fiasco.* Harmondsworth; Penguin

This is the most up-to-date book written on the topic by a journalist who followed the development closely and is a known adversary of the project. The book is intended to make a case against Concorde, but is highly informative and not purely polemical.

4 Costello, J. and Hughes, T. (1971). *The Battle for Concorde.* London; Compton Press

A very readable account of the full story.

5 Strang, W. J. (1971). 'Concorde.' *Proc. R. Instn Gt Br.* **44,** 218–238

Discourse by technical director of BAC. Interesting technical details and highly optimistic total commitment to the plane as a commercial and technical venture.

6 'But how many will fly in it?' *The Economist,* pp 85, 86, 11th December, 1971

Highly pessimistic appraisal of total sales prospects.

7 Maxwell-Hyslop, R. (1972). 'Concorde.' *Contemporary Review,* **221,** 17–22

Balanced political-economic view, more in favour than against, some technical detail.

8 'Three times dearer by Concorde.' *The Economist,* p 81, 8th June, 1974

Pessimistic assessment of the economics of Concorde.

9 'US critics firing again at Concorde.' *The Guardian,* 16th September, 1975

The saga over landing rights in the US continues.

10 Primack, J. and von Hippel, F. (1974). *Advice and Dissent.* New York; Basic Books

Inter alia an account of the political fight for and against the American SST

FURTHER READING

11 Adamo, J. G. U. and Haigh, N. (1972). 'Booming discorde.' *Geographical Mag.,* **44,** 663–664.

12 'Concorde – a special report.' *The Times,* 4th March, 1969

13 'Concorde – double your losses.' *The Economist,* **250,** 23rd March, 1974, pp 70–71

14 Stafford, E. M. and Ellison, A.P. (1972). 'Concorde cuckoo.' *New Society,* **8,** 514–515

15 'Five sold already', *The Economist,* **243,** 27th May, 1972, p 115

16 Walden, B. 'Flight of the white elephant.' *New Statesman,* 17th December, 1971, pp 847–848

17 West, R. 'Friends of Concorde.' *New Statesman,* 7th July, 1972, p 14

18 Maude, A. 'In defence of Concorde.' *Spectator,* 17th March, 1969, p 298

19 Brown, N. 'What future for Concorde?' *New Statesman,* **74,** 7th July 1967, pp 6–8

20 'Why Concorde failed.' *The Economist,* **213,** 14th November, 1964, pp 674–675

21 'I'm noisy, but still fly me,' *The Guardian,* 15th October, 1975

22 'Concorde, cynicism, and social technology.' Letters to the Editor, *The Guardian,* 20th October, 1975

23 'That discord again.' *New Statesman,* 11th March, 1977

24 'Survival of the fastest.' *The Guardian,* 27th October, 1976

Case Study Four
Thalidomide

The introduction of thalidomide onto the market and its use by thousands of people over several years, often with tragic consequences, is perhaps the most notorious example of a scientific advance gone wrong. The study of this case history can shed much light on the way the manufacture and use of drugs are controlled, on the way decisions on the marketing of drugs are taken and on many other aspects of the control and use of scientific discovery by society.

The thalidomide story has been written up by two Swedish authors, Henning Sjostrom and Robert Nilsson, and published as a Penguin Special in 1972 under the title *Thalidomide and the Power of the Drug Companies.* More recently (27th June, 1976) *The Sunday Times* published a long article on the subject, and a book by Harvey Teff and Colin Munro on the legal battles following the tragedy was published a short while ago. The damage caused by thalidomide had two separate aspects: one was the notorious and tragic damage caused to the fetus when thalidomide was taken during specific days of pregnancy; the other, less well known, was the often irreversible damage to the nervous system when thalidomide was taken for long periods of time.

The manufacturers of thalidomide, Chemie Grunenthal of Stolberg in West Germany, started manufacturing pharmaceuticals only in 1946 but grew fairly rapidly, though never to the status of a major manufacturer. Thalidomide, a new drug to be used as a sedative, came onto the German market in October 1957, under the name Contergan. During 1958 it was massively advertised and many companies in many countries started marketing thalidomide under licencing agreements. Soon thalidomide became widely used in a large number of countries under a bewildering array of trade names and in a bewildering number of combinations with other drugs (*see* page 39 of Sjostrom and Nilsson[2]).

Although several extravagant claims were made by the manufacturers for various therapeutic effects of thalidomide, essentially it was only used as a sedative and sleeping pill. Grunenthal, an entrepreneurial company with a small and inexperienced scientific staff, seemed to have made a discovery of major importance. Their new sedative drug Contergan (or thalidomide) appeared to have no acute toxic effects.

The most important question which arises out of the introduction of thalidomide is whether it was fully tested for its safety, in the light of knowledge and test procedures then available and in common practice, before it was put on the open market. In other words, were the manufacturers careless, negligent or merely unlucky. Sjostrom and

Nilsson deal with this question in Chapters 9 and 10 of their book. These authors seem to take the view that there was sufficient medical knowledge available to indicate the need for testing new drugs for possible fetal damage. They do concede, however, that such testing was not common practice for most drugs in most countries. Whether Grunenthal should have known the danger and tested the drug accordingly will, in our view, remain debatable. What is not debatable is that the drug was never permitted in the United States because the Food and Drug Administration was not convinced of its safety. Whether this was caused by better United States legislation, better administrative procedures, or merely by the outstanding ability of Dr. Kelsey of the F.D.A., is a matter of conjecture.

The second question is whether thalidomide was adequately tested even for use by people other than pregnant women. This question is also dealt with in Chapter 10 of the book by Sjostrom and Nilsson, and again these authors indicate that the tests were not carried out satisfactorily. It is left to the student to decide whether the argument is convincing. This will afford an opportunity to look into available methods of testing and think about the degree of possible elimination of risk.

One of the many disquieting features of the thalidomide story is the wide discrepancy in regulations between different countries. While thalidomide was never permitted on the United States market, it was available over the counter, without prescription, from every pharmacy in the German Federal Republic. The student might think about state control of the drug industry. He might also consider the need to take risks with drugs designed to cure serious disease as compared to the risks taken with drugs designed merely to alleviate discomfort.

Whatever one might think of the decision by those in charge of Chemie Grunenthal to introduce thalidomide onto the market, the decision to keep it on the market for as long as it was falls into a different class of error of judgment.

Thalidomide spread rapidly, under pressure from normal commercial advertising, because it was an effective sleeping pill and was apparently non-toxic, even in large overdoses. This property gave thalidomide a real advantage over other hypnotic drugs, notably barbiturates, which caused many deaths, presumably mostly suicides, by overdose. We must not fall into the trap of minimizing this advantage of thalidomide with the aid of hindsight.

The belief in non-toxicity did not last for long. Although no cases of acute poisoning were reported, the first reports of nerve damage by prolonged use of thalidomide began to appear fairly soon after its introduction. Polyneuritis, a nervous disease with symptoms such as numbness of hands and feet and, in more severe cases, lack of co-ordination in leg movement was observed. Chapters 3 and 4 of the book by Sjostrom and Nilsson give the detailed story of how reports of polyneuritis began to emerge and how the scientists and directors of Chemie Grunenthal fought to keep their drug on the market. The story makes sad but fascinating reading. Clearly the Grunenthal people did not want to believe that their drug was harmful. There is a world of difference, however, between just not believing a story and stubbornly clinging to one's own belief out of greed and self-interest. Were the Grunenthal scientists guilty or merely misguided and prejudiced? It is easy to judge, but difficult to do justice. It would certainly appear that Chemie Grunenthal were not only reluctant to accept adverse reports on their drug, but indulged in somewhat questionable techniques in defending their commercial interests. When, on the other hand, the damage to the fetus caused by thalidomide became clear, Grunenthal withdrew the drug fairly quickly from the market.

The issues

Amongst the many issues raised by the tragic thalidomide episode the student might perhaps consider the following.

1. Should the testing of new drugs be more tightly controlled by state agencies?
2. Is the rate of innovation in drugs too rapid (*see* [4])?
3. What were the decisions made in the thalidomide story and who made the decisions which we now see to have been wrong?
4. How was the scientific judgment of the protagonists in the struggle for survival of thalidomide clouded by their respective allegiances?
5. What is the relationship between legal truth and scientific truth (*see* Chapter 12 of [2])?
6. What are the main factors which caused the thalidomide tragedy?
7. Was the legal system able to cope adequately with the results of the tragedy (*see* [3])?

Bibliography and references

[1] Page, B., Knightley, B., Potter, E. and Terry, A. Thalidomide – the story they suppressed. *The Sunday Times,* 27th June, 1976

[2] Sjostrom, H. and Nilsson, R. (1972). *Thalidomide and the Power of the Drug Companies.* Harmondsworth; Penguin

[3] Teff, H. and Munro, C. (1976). *Thalidomide: The Legal Aftermath.* Saxon House

[4] Wibberley, D. G. (1975). 'Alternatives to new drugs.' *Pharm. J.* 24th May, p 462

Case Study Five
The Pollution of Water by Nutrients

Water may become seriously polluted when it carries an excess of the plant nutrients nitrogen and phosphorus. These nutrients can increase the biological productivity of the water, encouraging in particular the growth of algal 'blooms' on the surface. When the algae die they sink to the bottom where their decomposition can seriously deplete the quantity of oxygen dissolved in the water. This can eventually disrupt the whole ecology of the water system, economically important fish such as trout and salmon often being the first species to suffer. The problem is most severe in lakes where eutrophication can occur. The best known example of eutrophication caused in this way is Lake Erie, where valuable species of fish have disappeared, the water is cloudy with microorganisms, mats of algae cover acres of the surface and beaches and swimming are spoilt by decaying vegetation.

There are three major sources of this kind of pollution: sewage, agricultural run-off and detergents. If too much sewage is dumped in a given body of water, its decomposition can consume all of the water's oxygen with disastrous consequences. Primary and secondary treatment of sewage transforms its organic content into inorganic salts, including nitrates and phosphates. These make no immediate demands on the water's dissolved oxygen and so are often dumped without further treatment. In excess, however, these salts can stimulate the growth of algae and so lead, in this roundabout way, to oxygen depletion and eutrophication. The chief contributor of nitrogen to water in the United States is agricultural run-off which also adds further quantities of phosphorus. Fertilizer application has increased enormously in the past 25 years in those countries with advanced agriculture, with a corresponding increase in its associated pollution problems. A third source of phosphorus is detergents, which have largely replaced soap domestically and in industry. Less serious pollution problems have also been caused by agricultural run-off and detergents. Nitrates from the former have contaminated drinking water and earlier kinds of detergent caused unsightly foam on rivers and interfered with the normal operation of sewage works, many of which had to be modified to cope with the new chemicals.

We are clearly dealing here with unintended consequences of technology, but there are important differences in the way decisions have been made in the three cases. In the case of detergents there was a clear decision on the part of a handful of large companies to market the new products of their laboratories. There were also clear decisions about continuing sales in spite of the pollution problem, and about what resources to invest in the search for less damaging products. In many countries governments had to decide what action to take, if any, against the manufacturers. Underlying all these decisions are, of course, the decisions of millions of individuals to use detergent rather than soap. In the case of pollution by sewage effluent the problem seems to reduce largely to what people are prepared to pay for clean water, for nutrients can be removed with present technology. Here the decision-makers are all bodies finally answerable to government. The simplicity of the kind of decision involved here must not, however, be exaggerated, for every pound spent on clean water is a pound not spent on other desirable things, such as roads, hospitals, pensions and so on. In the case of fertilizer run-off, decisions were largely made by individual farmers choosing to apply more fertilizer to their land. There were, however, excellent economic reasons for them doing so, reasons generated by a whole network of decisions about such things as the price of fertilizer, the price of raw materials for fertilizer production, particularly energy, government agricultural policies, agricultural labour costs and so on. The worrying thing here is that most of these decisions were taken in complete ignorance of the pollution problem which they were to cause, for in most cases the pollution is a remote consequence of the decision, mediated by a whole network of other independent decisions. It is worth asking whether any decision-maker can be expected to deal with such a complex problem.

Most of the literature cited below refers to the situation in the United States. This is because the problems are at their most severe there and people are more aware of environmental issues.

The issues

The case study can be used to explore, among others, the following questions.

1. Has the problem of water pollution by nutrients, and in particular the problem of Lake Erie, been exaggerated, and, if so, why?
2. Could the damage to Lake Erie, arising as it does from so many causes, have been foreseen, and what steps might have been taken to limit the damage?
3. Why have fertilizer applications increased so much in recent years?
4. Pollution by fertilizer dissolved in agricultural run-off is a remote consequence of many independent decisions. Is it possible for such remote consequences to be considered by a decision-maker?

5. Why was the effect of detergent on sewage plant not investigated before the new products were marketed? How could the manufacturers have been persuaded to act on the findings of such an investigation?
6. What was the role of competition between rival manufacturers in the detergent controversy?
7. What problems face the introduction of new kinds of detergent which will do less damage to the environment?
8. Why has soap so largely been replaced by detergent in the home? Does this reflect a genuine advantage of detergent over soap, the availability of raw materials, or merely good advertising?
9. Many sewage works had to be modified in order to cope with the newly introduced detergents, the cost being met from public funds. Is this a satisfactory arrangement and, if not, what improvements can be suggested?
10. If human sewage could be returned directly to the soil there would be no problem of its disposal. Is this possible?

Bibliography

[1] Allen, H. and Kramer, J. (eds) (1972). *Nutrients in Natural Waters.* New York; Wiley

A collection of technical papers, many of which have extensive bibliographies. The most useful papers are likely to be those of Cohen and of Tenney *et al,* concerning the removal of nutrients from waste water, and that of Duthie, 'Detergent developments and their impact on water quality'.

[2] Beeton, A. (1969). 'Changes in the environment and biota of the Great Lakes.' In *Eutrophication: Causes, Consequences, Corrections.* National Academy of Science

A very good, technical account.

[3] Challis, E. (1974). 'The approach of industry to the assessment of environmental hazards.' *Proc. R. Soc.,* Series B, **185,** No. 1079, 183–198

How industry investigates the likely environmental impact of some of its proposed chemical products. Detergents are one of the examples used.

[4] Commoner, B. (1972). *The Closing Circle.* London; Cape, especially Chapters 4, 5 and 8

Commoner's thesis is that serious pollution is the result of a qualitative change in technology which began around 1945. Today the same basic needs are supplied as then, but by technologies which use artificial materials in place of naturally occurring ones.

These new technologies have, therefore, a much greater impact on the environment than the old ones. Fertilizer use falls into this pattern. Farmers are forced to use ever larger quantities of fertilizer, despite the environmental problems discussed in Chapters 4 and 5, nitrate contamination of drinking water and the eutrophication of lakes. The water pollution problems caused by sewage treatment and detergents are also discussed.

5 Commoner, B. (1968). 'The killing of a great lake.' In *The 1968 World Year Book.* Field Enterprises Educational Corporation

A detailed account of Lake Erie on which Chapter 5 of [4] above is based.

6 *Conference on Water Pollution as a World Problem* (1971). pp 101–140. London; Europa

Three papers and a discussion on the legal and political aspects of chemical and pesticide pollution of water, including pollution by nutrients.

7 Edmondson, W. (1971). 'Fresh water pollution.' In *Environment: Resources, Pollution and Society.* Ed. by Murdoch, W. pp 213–229. Stamford, Conn.; Sinauer

A very useful introductory overview of fresh water pollution, including eutrophication of lakes and its avoidance, agricultural drainage and the cases of Lakes Erie and Washington.

8 Goodman, G. (1974). 'How do chemical substances affect the environment?' *Proc. R. Soc.*, Series B, **185,** No. 1079, 127–148

An excellent account of the difficulties involved in determining whether a new chemical product will damage the environment.

9 Gunn, D. (1972). 'Dilemmas in conservation for applied biologists.' *Ann. appl. Biol.*, **72,** 105–127

Argues that many arguments from defenders of the environment are mistaken, unscientific and fly in the face of the facts. Nevertheless, these arguments are very popular and persistent. Many of the arguments surrounding Lake Erie fall into this category. A very useful article, with references.

10 Hynes, H. (1972). *The Biology of Polluted Waters,* 2nd ed. pp 140–145. Liverpool; Liverpool University Press

A brief discussion of the eutrophication of lakes and its acceleration by nutrients from sewage, agricultural waste etc., with many references to the technical literature.

11 Jeger Committee (1970). *Taken for Granted.* London; H.M. Stationery Office

The report of the Jeger Committee on Sewage Disposal, giving a readable background to the subject.

12 Jenkins, S. and Ives, K. (1973). *Progress in Water Technology. Volume 2: Phosphorus in Fresh Water and the Marine Environment.* Oxford; Pergamon

The report of the International Association of Water Pollution Research Conference in London in 1972. All aspects of pollution by phosphorus are considered, although all the papers are, of course, technical. Of particular interest are the papers of Devey and Harkness, Wood and Gibson, Jenkins *et al*, and Ashforth and Calvin.

13 Klein, L. (1960). *Aspects of River Pollution.* London; Butterworths

Includes an account of the 'Luton experiment' when biologically soft detergents were surreptitiously substituted for the, then, commonly used hard detergents in the town of Luton.

14 Marstrand, P. (1974). 'Assessing the intangibles in water pollution control.' *Int. J. envir. Studies* **5**, 289–298

In deciding how much we should pay for pure water, values must be placed on non-economic factors. This paper discusses various approaches to this problem.

15 McKnight, A., Marstrand, P. and Sinclair, T. (eds) (1974). *Environmental Pollution Control,* Chapters 5 and 6. London; Allen and Unwin

Chapter 5 gives a very useful overview of the problem of water pollution and its treatment. Chapter 6 deals with U.K. and international law on water pollution.

16 McL. Philp, J. (1974). 'A multi-national company, the public and the environment.' *Proc. R. Soc.* Series B, **185**, No. 1079, 199–208

How Unilever ensures that the marketing of its products will not represent a hazard to the consumer or damage the environment. One of their major products is detergent.

17 Milway, C. (ed) (1968). *Uppsala Symposium: Eutrophication in Large Lakes and Impoundments.* O.E.C.D.

The report of an international symposium. The most useful part of the work is likely to be the case studies it contains, although they are presented in much technical detail.

18 Newson, G. and Sherratt, J. Graham (1972). *Water Pollution.* Altrincham; Sherratt

A detailed, but not over-technical, guide for the layman, covering all legal aspects of water pollution in the U.K.

19 Nisbett, A. (1972). 'The myths of Lake Erie.' *New Scient.* 23rd March, 650–652

Argues that the commonly stated 'death' of the Lake is an illusion. It may be sick, but it is far from dead.

20 Prat, J. and Giraud, A. (1964). *The Pollution of Water by Detergents.* O.E.C.D.

A fairly technical, well-referenced, account of detergent pollution and possible counter-measures.

21 Report of the Study of Critical Environmental Problems (SCEP) (1970). *Man's Impact on the Global Environment,* pp 144–149 and 238–240. Cambridge, Mass.; M.I.T. Press

Deals with phosphorus in fertilizers, detergents and sewage effluent and its impact on the environment, especially the eutrophication of lakes. References to the technical literature are given, and a number of recommendations offered.

22 Southgate, B. (1969). *Water: Pollution and Conservation,* pp 67–77 and 124–136. Middlesex; Thunderbird

Pp 67–77 deal with the problems created for sewage treatment by the introduction into the U.K. of synthetic phosphate detergents. Pp 124–136 present a brief case study of the pollution problems, mainly of sewage, of the Thames estuary and their treatment.

23 Wagner, R. (1974). *Environment and Man,* 2nd ed., pp 110–121. New York; Norton

A brief, non-technical, account of eutrophication and pollution by sewage, detergents and agricultural run-off, with a discussion of some possible solutions and the case of Lake Erie.

24 Walker, C. (1971). *Environmental Pollution by Chemicals,* pp 77–80. London; Hutchinson

A brief account of some of the harmful effects of detergents on river life, water quality and sewage treatment and what has been done to mitigate some of the problems.

25 Zwick, D. and Benstock, M. (1971). *Water Wasteland.* New York; Grossman

Pp 67–91: a fascinating account of the debate over pollution by detergents and the various responses made by the United States Government and the detergent industry to the problem.

Pp 92–100: the scale of pollution by run-off from agricultural land and the dilatory way in which the problem has been tackled in the United States.

Case Study Six
Road and Rail Transport Policies

This case study might, most logically, deal with all forms of inland transport including canals, airways and pipelines. This would, however, make it very wide-ranging, so it has been limited to road and rail transport of goods and passengers. In other words, it is concerned with trains, buses, lorries and cars.

During the last 150 years, there have been two major changes in modes of transport on land; in the nineteenth century the change involved the widespread use of railways, in the twentieth century the change involved the widespread use of the internal combustion engine. In both cases there was a decline in the use of previously existing forms of land transport (coaches and horses, railways) and also a large increase in the total amount of movement of people and goods.

Although the details are different, a similar pattern can be seen in the two cases. In neither can it be said that a clear policy towards the new form of transport was adopted by national or local governments. In both industries, the early days are characterized by a large number of competing companies, some producing means of transport, others running transport businesses, and set up as people saw the financial possibilities of such ventures. Gradually, with financial collapse, mergers and take-overs, the number of companies declined until, in both cases, it reached some number under ten. Parallel to this was a gradual increasing intervention of the Government. For the railways this culminated in 1945 with nationalization, because they were in such financial difficulties. The relationship of the Government to road transport is more complex. The financing of road building has moved from the private turnpike trusts and parishes of the eighteenth and nineteenth centuries, to the present day local authorities and National Government. The same two agencies, local and National Government, have direct control over traffic and road regulations. The car industry, until very recently, has been entirely within the private sector; but this does not mean it has been beyond Government influence. The major interest of the Government in the car industry has been, and still is, an economic one. This interest developed most strongly just prior to, during and after World War II. Before and during the war the car industry and Government were involved in creating extra capacity for wartime engineering production; after the war the Government was concerned to persuade the car industry to produce a large proportion of cars for export. Since then, the car industry has continued to be a major employer and exporter; it is one of the most sensitive industries to economic recessions and is sometimes used by the Government to manipulate national earnings and expenditure.

Although Governments have had no clear transport policy in the form of an overall plan, it is possible to identify the basic approach that has been adopted towards the railways and the roads. Since 1945, the remit for British Rail has been to operate a commercially viable system, that is, one that financially breaks even, or better still, makes a profit. Policies, that would help to achieve this, have therefore been adopted. The result has been a very considerable cut-back of route mileage, stations, rolling stock and staff.

The attitude towards roads and traffic has been characterized by three major assumptions, as follows.

(1) That the car is a great asset and greatly improves the quality of life of those who have access to it; we look forward to the day when everyone has one.

(2) Everyone who wants a car has a right to own one and to drive it anywhere they please.

(3) The duty of National and local Government is to see that sufficient road space is provided to accommodate the current number of cars and the number predicted in the next few years.

Up until the 1960's, the only aspect of the car that was worrying and that did not quite fit in with these assumptions, was the accident rate. However, during the 1960's it began to be clear that cars in unlimited numbers in cities rather defeated the object of the exercise, because they caused such congestion. Buchanan, in the report 'Traffic in Towns[4], recognized this but inclined to the view that cities and their roads must be so designed to avoid it. In order to provide road space, particularly for the volume of traffic arriving in cities from the inter-city motorways, the obvious answer seemed to be urban motorways. But in the late 1960's and early 1970's public opinion moved clearly against them, on account of their destructive effect on the areas through which they were built and their cost. This is demonstrated by the Labour G.L.C. policy of rejecting the London motorway box plan and of winning the 1973 election, with that intention well publicized. Smaller scale attempts to provide enough urban road space are not on the whole successful – the more space, the more traffic appears to fill it. A rational solution to the problem in towns and cities might seem to be to restrict traffic access and to provide a well planned public transport system. However, in practice there are a great many difficulties with this approach and local government is very reluctant to adopt such schemes. Nottingham is a notable and interesting exception to this.

Meanwhile, the National Government continues to promote road transport by spending much larger sums of money on road construction and maintenance than is spent on British Rail investment. Admittedly, it can be argued that less needs to be spent on the railways because the major investment in them took place years ago, unlike the roads. However, they need continuous replacement and modernization of both line and rolling stock. The expenditure on these between 1962 and 1972 was well below replacement level. No doubt this was partly deliberate, in order to cut back the railways, but it seems to be partly the result of an attempt to make the books balance in the short term.

One of the major road expenditures is, predictably, on motorways. The stated policy of the Government towards motorways is as follows.

> The inter-urban road programme consists mainly of an extension to the existing network to create a national lorry network to minimise the environmental effects of heavy vehicles and avoid congestion in the interests of both road safety and economic efficiency.
> *(First Report from the (Parliamentary) Expenditure Committee on Transport 1974.)*

Over the last 20 years there has been a bigger change in the pattern of freight transport than there has in passenger transport. The change, of course, has been from the railways to the roads. However, there is little doubt that the railways produce less congestion and undesirable environmental effects than roads and lorries ever can. Even with inter-urban motorways lorries come off them somewhere and that is where the congestion occurs; they use more energy and produce more pollution per ton mile; roads occupy more land space than railways for the same carrying capacity.

Admittedly, there is more handling of freight on railways, if goods have to be transferred to lorries at some point; and freight depots can cause local congestion. With our current system of costing it is often cheaper to transport goods by road. Nevertheless, the road freight business runs at a profit, unlike the railways. However, as Mishan[10] points out in *The Costs of Economic Growth,* what is profitable and what is not depends on the costing and accounting system that is used. For example, road users do not pay directly for the cost of road accidents; this is paid for by the whole community via taxes and rates and is greatly in excess of the cost of rail accidents. No costing is done at all of the effects of pollution, noise, stress, alternative land use and so on.

Overall, in terms of the criteria mentioned in the quote above, it is very doubtful that the right balance has been struck between road and rail transport. The attitude of the Government and planners alike seems to be still very largely based on the three assumptions listed earlier, even though there is increasing evidence to suggest that a different attitude may be called for in different places and times. For example, the attitude to the car in cities may need to be very different from the attitude towards it in rural areas.

There are three reasons worth mentioning that may partly explain this reluctance to adopt a more flexible approach to the place of road transport. (Their order of listing does not indicate order of importance.)

First, there is a well organized, well financed road lobby, which is continually pressing the case of the road users, whilst there is no equivalent for the railway. This lobby attempts to influence both local and National Government and, like any other lobby, has three main ways of doing this: (1) direct contact with relevant Government departments; (2) political pressure by attempting to influence M.P.s or councillors; (3) public campaigns to influence voters.

The road lobby is able to operate mostly through the first of these; rarely is it involved in a public campaign. This, of course, is why the public know very little about the lobby. However, it should be noted that it is generally accepted in the corridors of power that the power of a lobby is inversely proportional to its publicity.

Secondly, as already mentioned, the road vehicle industry is important economically, as are the associated industries of petrol and road building. It is possible, of course, to change the nature of the product, using the same workforce and much of the same capital equipment. But this would undoubtedly involve some major re-thinking and re-organization.

Thirdly, the dominant ideology in Britain is one which places a high priority on individual freedom and which disapproves of extensive Government control. In relation to transport, this means that any restrictions on the private use of cars or lorries goes against the grain. This argument often ignores the fact that not everyone has access to a car, and people who do not have such access are often adversely affected by those who do. The adverse effects take the form of environmental degradation, accidents (the majority of which, in urban areas, involve pedestrians) and the erosion of public transport services. In this context, it is worth noting that only just over 50 per cent of households own a car and that the women, children, elderly and infirm often have limited or non-existent access to it. Although more households could become car owners, this would increase congestion problems and there will always be a considerable proportion of the population unable to drive a car. It is to some extent one man's freedom, another man's non-freedom.

This case study therefore raises two major issues, which are to some extent separate and to some extent linked. One is concerned with finding the 'best' balance between rail and road transport, particularly of freight. The second is concerned with finding the 'best' balance between private and public transport for people. They are linked because they make use of the same road and rail facilities and because the quality of the environment is affected by the transport of both goods and people.

The issues

This case study may be used to consider, amongst others, the following issues.

1. What should be the aims of a national transport system? To what extent should it be based on forecasting and what extent on planning? How would it deal with the different needs of urban, rural, and long distance transport? How would it promote mobility in a socially equitable manner?
2. Which aspects of this policy would be best dealt with by National Government and which by local government?
3. A major problem that any planned policy faces is that whilst plans have to be made by a relatively centralized body, transport decisions are made by individuals. How could the following measures be used effectively and fairly to aid the implementation of a plan:

monetary measures	e.g. a road tax, bus fares.
restrictions	e.g. restricted access to areas.
persuasion	
availability of resources	e.g. buses, road space?

4. Which groups of people are affected by the various aspects of a transport policy? How are they consulted or involved in the decision-making process?
5. How much are problems to do with new uses of old technology and how much are they to do with designing new technology?
6. Which of the external costs of railways and roads could and should be costed in numerical terms? How can we deal with those externalities which are exceedingly difficult, or impossible, to evaluate in numerical terms?
7. Does increased mobility necessarily mean an increase in the quality of life?

Bibliography and references

1 Bagwell, P. S. (1974). *Transport Revolution From 1770.* London; Batsford

Transport changes in Britain from 1770 to 1970.

2 Dyos, H. J. and Aldcroft, D. H. (1969). *British Transport.* Leicester Univ. Press

Economic survey of all transport forms from the seventeenth century to 1939.

A wide range of publications, dealing with greater historical detail, can be found in the bibliographies of the above two general histories.

3 *Cars in Cities* (1967). London; Ministry of Transport
A study of trends in the design of vehicles with particular reference to their use in towns.

4 *Traffic in Towns* (the Buchanan Report) (1963). London; Ministry of Transport

5 Cross and Tetlow (1968). *Homes, Town and Traffic.* London; Faber

6 *The Board's Review of Railway Policy* (1973). British Railway Board
Most recent proposals for rail plans up to 1981.

7 Plowden, W. *The Motorcar and Politics in Britain.* London; Bodley Head (1971): Harmondsworth; Pelican (1973)
Account of public policy towards the car in Britain from 1890 to 1970.

8 *Changing Direction* (1974). Independent Commission on Transport. London; Coronet Books
An investigation into Britain's transport facilities and policies with particular emphasis on their effect on people and environment. Contains some recommendations for the future.

9 Hillman, M., Henderson, I. and Whalley, A. (1973). *Personal Mobility and Transport Policy.* Political and Economic Planning (PEP) Broadsheet 542
Investigation of personal mobility from the points of view of the individuals, the community, and policies affecting how people get about.

10 Mishan, E. *The Costs of Economic Growth.* London; Staples Press (1967): Harmondsworth; Pelican (1969)
Critically examines the effect of continued economic growth on our lives. Two chapters consider the effect of growth on cities and transport.

11 Thomson, J. M. (1974). *Modern Transport Economics.* Harmondsworth; Penguin
Applies an economic analysis to all forms of transport – rail, road, sea and air. Includes basic economic facts of transport, problems that arise in attempting to achieve a balance between supply and demand and ways of dealing with these problems.

12 Nove, A. (1973). *Efficiency Criteria for Nationalised Industries.* London; Allen & Unwin
Concerned with the criteria used to judge the efficiency of nationalized industry and with the oversimplified unrealistic economic models still used in economic textbooks.

13 Ogilvie, J. and Johnson, B. (1975). 'Railways: accounting for disaster.' *New Scient.*, 23rd October
Concerned with the problem of calculating reasonable subsidies for road and rail in the absence of an adequate costing mechanism for

their social, environmental and economic usefulness. Also concerned with overmanning on the railways.

14 Bendixson, T. (1974). *In Place of Cars.* Temple-Smith
Takes the view that we need to change the way we move things around, but that there is no single panacea. New city planning, new uses of old technology, new technology.

15 Hamer, M. (1974). *Wheels within Wheels.* Friends of the Earth
Analysis of the road lobby – what it is, how it works and the kind of influence it may have.

16 Barker, A. and Rush, M. (1970). *The M.P. and His Information.* London; Allen and Unwin
For chapter by W. Plowden on the roads lobby.

17 Kimber, R. and Richardson, J. J. (eds) (1974). *Campaigning for the Environment.* London; Routledge and Kegan Paul
Collection of papers including several to do with transport decisions and public reaction.

18 *Energy in Transport, British Rail 1975–2000, An Electrifying Case, The Nottingham Transport Policy.* London; Transport 2000 Publications
Transport 2000 is a widely based independent group concerned with transport and its impact on society. It produces reports and recommendations, like the four above, has 22 affiliated national organization and 25 local groups.

There are a great many publications, relevant to this study, available from the Government and other interested organizations. A full list of these sources would be very long; a small sample is listed below:

Parliamentary Expenditure Committee Reports – H.M. Stationery Office e.g. Public Transport 1974, Urban Transport Planning
White Papers – H.M. Stationery Office
Department of Environment Reports
H.M. Stationery Office Reference Pamphlets
Transport and Road Research Laboratory – H.M. Stationery Office
Chartered Institute of Transport
British Road Federation – especially British Road Statistics
British Rail
Local Authorities (for local transport problem and schemes)
The three Parliamentary Parties
Journal – *Transport Economics and Policy.*
Published 3 times a year. Fairly advanced and specific articles.

Case Study Seven
The Aswan High Dam

The River Nile is known to have been a cause for concern to the Egyptians for many centuries. The difficulties arise because Egypt is dependent on the Nile for its source of irrigation water but the flow rate of the river fluctuates enormously during each year. This results in flooding, which can be quite severe and damaging to land and villages, during August, September, October and near drought in April, May and June. Using the natural flow of the river makes it possible to grow only one crop a year on 4 per cent of the land area.

Early in the nineteenth century, one of the Governors of Egypt began building a series of barrages, not to store water so much as to raise the level of the river, in order to use the canal system to irrigate the land all the year round.

Barrages continued to be built and enlarged until 1938. In 1902 the first Aswan Dam was built and functioned, not to raise the level of the river, but to store water when the river was in flood for the rest of the year. This enabled a larger area of land to be brought into agricultural production for more of the year.

However, it was not big enough to store water for more than a year; it could not supplement the supply of water in a year of low flood nor could it entirely prevent damage to the farmlands in periods of excessive flood. Added to this, Egypt's population was increasing and more agricultural land was urgently needed.

Year	*Population*
1800	2.5m
1882	7 m
1917	13 m
1930	17 m
1955	23 m
1965	30 m
1975	37 m

During the late 1940's, two plans were drawn up to increase water storage capacity along the Nile. One plan, devised by the Egyptian Irrigation Service, involved using the great African lakes, Albert and Victoria, for water storage, by building a series of dams and a canal. The other plan devised by an Egyptian engineer Daninos, involved building a High Dam, just up river from the original Aswan Dam, big enough to provide a multi-year storage lake and considerable hydro-electric power.

In 1952, after the overthrow of King Farouk, Nasser and the military government decided to go ahead with the Aswan High Dam. In fact, the building of this enormous structure became the cornerstone of the country's long-term planning and modernization program. The major reasons for this choice would seem to be as follows.

(1) The scheme involving the great lakes would take at least 25 years to complete whilst the Aswan High Dam would take about 6 years.

(2) Most of the great lakes scheme would be situated outside Egypt in other African countries, whereas the Aswan High Dam would be in Egypt. This means the latter would be entirely under the control of the Egyptian Government which could use it to plan, in advance, the crop system.

Thus, in 1952 the West German Government was asked to help draw up the details of the plan. The dam was to cost $1,000m, an amount of money Egypt could not possibly raise on her own. Nasser knew that the U.S.S.R. was interested in the project and might be prepared to give or lend some money for it. But he preferred the idea of aid from the West. The plan was basically approved in 1955 by the World Bank, the United States and the U.K. and between them they agreed to contribute $400m. By 1956, the United States Government had become unsympathetic towards Egypt and the plan for a number of reasons.

(1) Egypt was continuing to be antagonistic towards Israel.

(2) Building the dam would strengthen Arab nationalism which was not in the interests of the United States.

(3) Building the dam might improve Egypt's cotton harvest, which was in competition with the United States cotton production.

(4) Egypt had bought arms from the U.S.S.R.

(5) Egypt recognized Red China.

As a result the United States not only withdrew its own offer of aid, but also put pressure on the World Bank to do likewise. This is one of the clearest examples of the World Bank yielding to political pressure from its most powerful member. Britain of course followed suit.

Egypt then turned to the U.S.S.R. and arranged to borrow some of the money at a low rate of interest and to use U.S.S.R. equipment and engineering assistance. Some of the rest of the money was raised by the Egyptian nationalization of the Suez Canal, which gave Egypt access to the revenue which had been collected by the Anglo-French Suez Canal Company.

In 1960, some years later than originally planned, work on the dam began. By 1970 the major work of building the dam was completed. 500,000 acres of land had been reclaimed or developed by 1965 and a further 1m acres were expected to be reclaimed or developed during the 1970's. This represents an increased useful land area of 30 per cent. In addition, the dam contains twelve turbines which generate 10 billion kW hours per year whilst 6 billion are generated from other sources. Even so, this is not enough electricity to supply power to all the villages and to heavy industry.

Also by 1972, a number of other consequences of building the dam were beginning to be apparent. These result from the altered hydrology and ecology of the area, as less water and other materials flow down the river and as greater areas of land are permanently irrigated.

Two major differences between the first Aswan Dam and the Aswan High Dam need to be noted.

First, and quite obviously, is their difference in size. The Aswan High Dam contains 34 times as much material as the old Aswan Dam; Lake Nasser is therefore much bigger than the reservoir behind the first Aswan Dam.

Secondly, the old Aswan Dam had sluices that were opened after each storage period to allow it to drain. There was a considerable downfall of water which carried with it most of the silt that had collected behind the dam. The Aswan High Dam does not have such a system as it is designed to store water from one year to the next. Even if it had similar sluices, it is most unlikely that much of the silt would be carried away by the water, because turbulence is relatively local compared to the size of the lake. All the silt that arrives in Lake Nasser will accumulate there. Large dams of this kind are estimated to have a life of 70 years. At the end of that time, the water has to be drained out and either the reservoir dredged or the dam broken up.

THE UNDESIRABLE OUTCOMES OF BUILDING THE ASWAN HIGH DAM

(1) *The lack of silt deposition on agricultural land.* Before the Aswan High Dam was built there was annual flooding of the agricultural land on either side of the Nile. The water carried silt and mineral salts, brought down from higher up the river in Ethiopia. This provided annual fertilization of the land, which no longer happens. The soil is beginning to be depleted of a number of elements; it is estimated that N, P, Ca, Mg will be almost absent from the soil in 4 years, and Cu, Zn, Mg, B, Mn in 15 years.

(2) *River erosion.* The silt-free water from the dam is eroding the river banks, barrages and bridges much more quickly than the silty water did.

(3) *Coastal erosion.* The coast line of the Nile Delta is retreating inland, due to the westward drift of the sea. This used to be counter-balanced by the silt brought down by the Nile; there is now no counter-balancing.

(4) *Reduced fish catch.* The fish catch in the E. Mediterranean and in the Nile delta lakes has dropped significantly since the building of the Aswan High Dam. In 1962 in the E. Mediterranean the total fish catch was 30,600 metric tons. In 1968 this was down by nearly 60 per cent due to the almost total elimination of sardines. Similar trends are observable in the delta lakes. This is because there is no annual

flood water or silt; the flood water flushed out the dirty salty water from the delta and sea and the silt carried nutrient chemicals with it.

This is to some extent offset by the potential of Lake Nasser as a source of fish. It is expected that 10,000 metric tons of fish could be caught annually from the Lake.

(5) *Increase in bilharzia.* Bilharzia is a disease common in many parts of Africa, including Egypt. It is transmitted by a species of snail. During the dry season many of these snails used to die but now there is no dry season, this means of controlling their population has gone. The number of snails and the incidence of the disease have significantly increased.

(6) *Increased mosquitoes and weeds.* For the same reason as the increase in snails, there has also been an increase in mosquitoes, some of which carry malaria, and in weeds on agricultural land.

(7) *Water loss.* Considerable quantities of water are lost from Lake Nasser by evaporation and seepage. An estimate for 1971 when the lake was half filled (11,000m cubic metres) was 10m cubic metres.

Most of these consequences of building the Aswan High Dam were predicted by the Chairman of the Egyptian Hydro-electric Commission, Dr. Abdel Aziz Ahmed, and others. The main unexpected outcome was the serious erosion of the banks of the Nile. Dr. Ahmed favoured the alternative scheme of using the African lakes, on the basis that less water would be lost from them, than from Lake Nasser, and there would be less disruption of the hydrological and ecological system of the Nile. However, for reasons already given, the Aswan High Dam was built.

The issues

This case study can be used to consider, among others, the following issues.

1. What aims did the Egyptian Government have in building the dam? What was the order of importance attached to these aims by the Egyptian Government?
2. What action can Egypt take to ameliorate some of the worst effects of building the dam?
   - lack of silt
   - reduced fish catch
   - bilharzia
3. On the basis of the data given, was the decision overall a good one? What further data would you need to be more certain of your judgment? What hidden costs can you identify?

Bibliography

1 Little, T. (1967). *Modern Egypt.* London; Benn
History of Egypt from ancient times to 1965. Chapter 16 particularly useful in relation to the Aswan High Dam.

2 Little, T. (1965). *The High Dam at Aswan.* London; Methuen
Account of the origin of the plans, the raising of the money and the actual building of the dam. Includes information on the alternative , ut forward by Dr. Abdel Ahmed. Barely mentions side-effects.

3 Nutting, A. (1972). *Nasser.* London; Constable
An account of Nasser's modernization programme for Egypt, internal policies and external international relations. Written by a political diplomat.

4 Farvar, M. T. and Milton, J. P. (eds) (1973). *The Careless Technology.* London; Tom Stacey
A collection of specialist papers on the unintended and undesirable consequences of technology in the middle to low industrialized countries. Includes three papers (Nos. 10, 11, 12) on the effects of the Aswan High Dam. Whole book is a very useful collection.

5 Mountjoy, A. (1972). 'Egypt cultivates her deserts.' *Geographical Magazine,* January, 1972
Description of the major policies of land reform, land reclamation and industrial investment. No mention of undesirable consequences.

6 'A blessing or a curse'. In *The Economist,* May 15th, 1971
Concerned mostly with undesirable consequences. Maintains that most of them were foreseen before the dam was built.

7 *The Middle East and North Africe 1975–76.* London; Europa
A survey and reference book for all the Middle Eastern and North African countries. Particularly useful for historical and economic information. Documents positive outcomes of building the Aswan High Dam but hardly mentions undesirable ones.

8 Allaby, M. (1974). *Who Will Eat*? Dawlish, Devon; David and Charles
Concerned with food production and distribution in the world. Includes a section on the building of the Aswan High Dam which highlights the undesirable consequences rather more than the desirable.

9 Mansfield, P. (1973). *Nasser's Egypt.* Harmondsworth; Penguin

OTHER LARGE DAMS

10 Russell, H. Bernard and Pelto, P. J. (eds) (1972). *Technology and Social Change.* London; Collier-Macmillan

'Kariba Dam Project', Scudder, T. and Colson, E. Very good detailed account of the social, economic and agricultural consequences of building this dam. Useful because these aspects are not covered in detail, in the bibliography for the Aswan High Dam.

11 Farvar, M. T. and Milton, J. P. (eds) (1973). *The Careless Technology*. London; Tom Stacey

Contains one paper (No. 13) on some of the effects of the Kariba Dam.

Case Study Eight
Car Production Technology

The car industry is an example of industry that uses large-scale elaborate technology in the production process. The technology was developed, of course, for economic reasons, but it also has a very marked effect on the lives of those people who work in the industry. It is the latter that this case study is concerned with.

We can identify three styles of production process that have arisen in a historical sequence, in the car industry, as well as in production generally: craft production, mechanization, automation. This does not mean that all production will eventually be automated; for one reason or another some parts of processes may not be suited to it.

The first cars to be produced were made in craft workshops where much of the work was labour intensive and highly skilled. They were made in small numbers at prices that could only be afforded by the wealthy.

It was Henry Ford who, around the turn of the century, saw the possibility of producing a great many more cars much more cheaply, for a mass market. He organized production such that each task was divided into as many small, simple operations as possible. Each man did one of these operations and the car parts were moved from worker to worker by machinery. Where feasible, machinery was also used to aid the completion of each operation in as short a time as possible. Hence, this change was known as mechanization.

The jobs created by this method were highly efficient but also highly repetitive and boring, with the same movements being repeated several hundred times a day. This fact did not escape Henry Ford's attention; however, he believed that while some people, himself included, would not be able to stand such an occupation, others were well suited to it, or were prepared to do it if compensated sufficiently by monetary rewards.

The third and most recent style of production is automation; this involves machines which are programmed to perform a number of tasks, and which may have some built in feedback mechanisms, but which are monitored and adjusted by workers. The latter receive information, usually in a centralized place, about what is happening through the use of measuring instruments. Where necessary they adjust the process, usually also from a centralized position.

The car industry uses automation at the moment only to a limited extent; of the identifiable processes – casting, forging, machining, pressing, welding, assembly, painting – machining is almost fully automated whilst welding has a considerable degree of automation and all the rest are mechanized, but not automated.

Although the type of production technology is the major aspect of industry that affects the lives of those who work in it, there are three other associated aspects which must be mentioned.

First is the division of labour, because, even using the same production technology, it may be possible to allocate tasks to people in a variety of ways. For the most part in the motor industry, there is a very high division of labour, this being considered the most efficient arrangement. The average time span of a task on the assembly line, for example, is one minute. Another aspect of the division of labour is the way in which decision-making and brainwork is almost totally removed from the shop-floor worker to the province of the owners, management and other professional groups such as design engineers, work-study people and personnel managers.

Second is the style of organization in the industry. This refers to the way that employers and employees relate to each other. Their interactions may be guided by custom, old practices and special personal loyalties between the two; or they may be guided by a set of formal rules and procedures, which tend to ignore personal considerations. The latter style is typical of many of the contemporary bureaucracies to be found in industry, government and social services. It is this style which operates in the car industry.

Thirdly, the economic structure of the industry has a marked effect on the people who work in it. This includes economic factors such as growth rates, trends in demand, product competition, profit margins, relative costs of capital investment, materials and labour. Car firms vary in their economic characteristics, depending partly on whether they are mass-producing cheaper cars or making a smaller number of quality cars. Until recently the car industry has been a growth industry, but has for a long time been subject to seasonal fluctuation in demand, especially for the cheaper cars. Because the cheaper cars are sold with a relatively small profit margin, on the basis of mass sales, and because labour costs represent a fair percentage of production costs, employment has also fluctuated seasonally. When the market is good, work is speeded up and overtime is compulsory; when the market is bad there is no overtime, workers are laid off or put on a short week. This kind of fluctuation does not apply so much to the quality car production, as demand is more stable and profit margins higher.

The effects of these characteristics of the motor industry on the people who work on the shop floor are well documented and analysed. The most obvious of these are the lack of control that workers have over the jobs they do, the nature of the product, the kind of production technology used, the composition of tasks, the rate at which they work and so on. And then, because each person only does a very small part of the overall process and each item is exactly the same as all the others for a particular model, there can be no identification by the workers with their product. This makes their work almost totally meaningless and unfulfilling. In certain industries powerlessness and meaninglessness are somewhat alleviated by a sense of belonging to the social group there. However, it is difficult to achieve a sense of belonging to a car factory. They tend to be very large places in which there are few stable work groups, workers are often moved within the plant or leave, and the style of management is bureaucratic. Furthermore, there is a compressed wage and skill distribution, which affords little opportunity for promotion.

The overall effect of these aspects of work in the car industry is that most workers experience no involvement or self-fulfillment in their work. This produces a kind of depersonalized detachment in which they continue to do the job to get the money at the end of the week, but they are being used by the employers, and in a sense by themselves, as things. This self-estrangement is what is meant by alienation, a concept first well developed by Marx. Marx emphasized a somewhat different aspect of alienation, its root causes rather than its socio-psychological manifestations. He maintained that alienation arises from the basic economic arrangement of production, whereby workers own neither the means of production nor the product of their work. Instead, both of these are owned by the capitalist minority; thus they also control the supply and nature of jobs, the supply and nature of the products. The majority of people are then in the position of having to sell their labour, and do jobs which are in no way creative or fulfilling; work becomes merely a means of sustaining life. For Marx this is a fundamental contradiction of the nature of man. A further such contradiction arises, according to Marx, because people no longer meet to exchange gifts that are an expression of themselves; instead they meet to exchange commodities which are in no way an expression of themselves. People thus become merely functionaries who manipulate money and objects against other people.

All these aspects of alienation have appeared in their most extreme form relatively recently in capitalist industrial societies. In many non-industrial societies the feudal pattern of land ownership is the basis of a somewhat lower level of alienation. It is only in societies with no private ownership, no substantial division of labour and no commodity production that alienation is non-existent.

Clearly, the incidence of alienation is worrying mainly because it raises a moral question. Is it right to perpetuate a system that produces such a high level of alienation, affecting the basic feelings of identity and self-esteem, of the vast majority of people? It also gives rise to practical problems regarding the organization of industrial production. There are increasing signs of dissatisfaction and frustration amongst the car workers. Absentee rates are generally high, quality of work is often poor to the point of 'rogue' cars being produced, workers are generally uncooperative and bloody-minded and strikes are common. The cause of strikes is complex and certainly is not only to do with the style of production. Reading what car workers themselves say about their work makes it quite clear, however, that style of production has a major influence on their attitude towards work.

It can be argued, of course, that if people are still prepared to do the jobs they cannot be that bad. However, when people need the money and have few job choices, they are prepared to do most unpleasant tasks. Often, people taking jobs in the car industry only take them 'temporarily'. Once there, various economic and psychological processes come in to play, such that many of them come to terms with the work. This may involve compromising drastically with their aspirations, maintaining a hostile attitude to work and accepting that most of their social and affective needs will be satisfied by their family relationships, not at work.

At any rate, car production is sufficiently disrupted by these labour-relation problems for several companies to be trying out alternative styles of production. Renault has tried three types of modification: job rotation in which workers move on to the next job at the end of each hour; job enlargement in which each worker does several operations in a sequence instead of only one; and job enrichment in which the workers work in teams, off the assembly line, but at a pre-set hourly rate. The last of these is the one most preferred by the workers.

Volvo, in Sweden, have such high rates of absenteeism and turnover of workers that they have built a new plant, designed jointly by workers and management, for all the production process to be carried out by teams of 15–20 people each. This may actually cut the absentee rate, but even if it does not do this, it enables car production to continue despite a high rate of absenteeism. This is to be contrasted with the traditional process, which cannot continue unless almost all the stations are manned.

Ford, on the other hand, maintain they cannot move to team production, because it would not be as efficient as the current system and they, unlike Volvo, depend upon mass production and sales. Ford, therefore, plan job enlargement, as in the Renault works, and a program of full automation, where this is feasible and economic. This would create a very different style of job, and, of course, fewer of them. Thus, a cluster of issues arise concerning the kind of production technology used, the type of jobs created, the number of jobs created, the distribution of these jobs and, perhaps above all, who should make the decisions that so fundamentally affect the lives of working people. This means considering the roles of management and shareholders of local industry and, increasingly, multi-national companies, the workers in these industries, National Government and now the E.E.C. At the moment major decisions are made very largely according to economic criteria by directors, managers and Government. In multi-national companies decisions are often made by people in a different country from the one where working people are affected. This raises enormous and complex questions about the ownership, control and direction of industrial enterprise all over the world. This case study does not attempt to examine these questions in their entirety, rather it examines just one of the undesirable/controversial outcomes that follow from contemporary industrial activity.

The issues

This case study can be used to consider, among others, the following issues.

1. How is – job rotation
   – job enlargement
   – job enrichment
   – automation

   likely to effect the forms of alienation listed by Blauner and the more fundamental cause of alienation identified by Marx?
2. How much is alienation due to the kind of technology used and how much to the socio-economic organization of production?
3. What are the merits and demerits of various combinations of – labour intensive and capital intensive work?
   – employment, unemployment, work sharing?
4. What does worker control offer in terms of improving work satisfaction, decreasing alienation and decreasing the division of labour, especially between manual and mental work?
5. If the family is the only place where social and affective needs can be satisfied, will it be under undue strain?
6. Can creative leisure activities compensate for non-creative work?

Bibliography

1 Blauner, R. (1964). *Alienation and Freedom.* Chicago; University of Chicago Press
Essential book containing a comparison of alienation and its causes in four industries with contrasting production styles: printing, textiles, car production, chemical plant.

2 Braverman, H. (1974). *Labour and Monopoly Capital.* Monthly Review Press
Very readable Marxist analysis, from both a historical and contemporary perspective, of changes in the production

process. It covers the rise of the factory system, mechanization, mass production and computerization. These changes are related to the needs of the owners of capital and their implications for workers are examined.

3 Fischer, E. and Marek, F. (eds) (1970). *Marx in His Own Words.* Harmondsworth; Penguin
Brief guide to what Marx said, written partly in Marx's own words and partly in the words of the editors. See especially Chapter 3, 'Division of Labour and Alienation'.

4 Goldthorpe, J. *et al.* (1969). *The Affluent Worker in the Class Structure.* Cambridge; Cambridge University Press
Survey of attitudes of well paid workers in three different industries in Luton (U.K.) – ball-bearing factory, car factory and chemical works. Attitudes to work are related to the social life and aspirations of workers.

5 Beyman, H. (1973). *Working for Ford.* Harmondsworth; Penguin
Description of what it is like to work in a car factory (U.K.) often in the words of the workers themselves. Raises major issues about work and industrial relations.

6 Hartley, J. (1974). 'Work-how-you-like modules.' *Engineer,* 25th July
Brief description of the modifications Renault are making to production style and labour relations.

7 Gyllenhammer, P. (1974). 'Changing work organisation at Volvo.' In *Industrial Participation.* London; Spring
Brief description of Volvo's attempts to improve job quality – notably at their new plant designed for team work.

8 Callahan, J. M. (1974). 'Volvo – flexible production.' In *Automotive Industries.* 1st October
Brief description of Volvo's new plant designed for team work. Contains more detail of layout and operation than article by P. Gyllenhammer, above.

9 *The Future of Work,* Units 10–11 of the Open University Course, Man Made Futures. Open University Press, 1975
Contains a reprint of [8], plus many other useful references and ideas.

10 Douglas, J. (1974). 'Assembly line stays – but £5000 a year men may be running it.' *Engineer,* 25th July
Brief account of changes in the production process at Ford U.K. plants, notably job enlargement and automation.

11 Coates, K. and Topham, T. (1974). *The New Unionism – The Case for Workers' Control.* Harmondsworth; Penguin
Maintains that full self-management is the only way of attaining a rational and humane organization of work. Examines opportunities and dangers.

12 *British Leyland — The Beginning of The End?* Counter Information Service
A non-establishment but straight account of the activities and implications of British Leyland both in the U.K. and abroad.
13 Hailey, A. (1971). *Wheels.* London; Michael Joseph
Vivid novel about life, at all levels, in the Detroit car industry.